Sebastian Mieruch

# Analysis of global water vapour trends based on satellite data

Sebastian Mieruch

# Analysis of global water vapour trends based on satellite data

## Statistical identification, error and significance estimation, comparative and correlative analyses

Südwestdeutscher Verlag für Hochschulschriften

**Impressum/Imprint (nur für Deutschland/ only for Germany)**
Bibliografische Information der Deutschen Nationalbibliothek: Die Deutsche Nationalbibliothek verzeichnet diese Publikation in der Deutschen Nationalbibliografie; detaillierte bibliografische Daten sind im Internet über http://dnb.d-nb.de abrufbar.

Alle in diesem Buch genannten Marken und Produktnamen unterliegen warenzeichen-, marken- oder patentrechtlichem Schutz bzw. sind Warenzeichen oder eingetragene Warenzeichen der jeweiligen Inhaber. Die Wiedergabe von Marken, Produktnamen, Gebrauchsnamen, Handelsnamen, Warenbezeichnungen u.s.w. in diesem Werk berechtigt auch ohne besondere Kennzeichnung nicht zu der Annahme, dass solche Namen im Sinne der Warenzeichen- und Markenschutzgesetzgebung als frei zu betrachten wären und daher von jedermann benutzt werden dürften.

Verlag: Südwestdeutscher Verlag für Hochschulschriften Aktiengesellschaft & Co. KG
Dudweiler Landstr. 99, 66123 Saarbrücken, Deutschland
Telefon +49 681 37 20 271-1, Telefax +49 681 37 20 271-0
Email: info@svh-verlag.de
Zugl.: Bremen, Universität Bremen, Diss., 2009

Herstellung in Deutschland:
Schaltungsdienst Lange o.H.G., Berlin
Books on Demand GmbH, Norderstedt
Reha GmbH, Saarbrücken
Amazon Distribution GmbH, Leipzig
ISBN: 978-3-8381-1365-4

**Imprint (only for USA, GB)**
Bibliographic information published by the Deutsche Nationalbibliothek: The Deutsche Nationalbibliothek lists this publication in the Deutsche Nationalbibliografie; detailed bibliographic data are available in the Internet at http://dnb.d-nb.de.

Any brand names and product names mentioned in this book are subject to trademark, brand or patent protection and are trademarks or registered trademarks of their respective holders. The use of brand names, product names, common names, trade names, product descriptions etc. even without a particular marking in this works is in no way to be construed to mean that such names may be regarded as unrestricted in respect of trademark and brand protection legislation and could thus be used by anyone.

Publisher: Südwestdeutscher Verlag für Hochschulschriften Aktiengesellschaft & Co. KG
Dudweiler Landstr. 99, 66123 Saarbrücken, Germany
Phone +49 681 37 20 271-1, Fax +49 681 37 20 271-0
Email: info@svh-verlag.de

Printed in the U.S.A.
Printed in the U.K. by (see last page)
ISBN: 978-3-8381-1365-4

Copyright © 2010 by the author and Südwestdeutscher Verlag für Hochschulschriften Aktiengesellschaft & Co. KG and licensors
All rights reserved. Saarbrücken 2010

# Abstract

Global water vapour total column amounts have been retrieved from spectral data provided by the Global Ozone Monitoring Experiment (GOME) flying on ERS-2, which was launched in April 1995, and the SCanning Imaging Absorption spectroMeter for Atmospheric CHartographY (SCIAMACHY) onboard ENVISAT launched in March 2002. For this purpose the Air Mass Corrected Differential Optical Absorption Spectroscopy (AMC-DOAS) approach has been used. The combination of the data from both instruments provides a long-term global data set spanning more than 12 years with the potential of extension up to 2020 by GOME-2 data on MetOp.

Using linear and non-linear methods from time series analysis and standard statistics the trends of water vapour columns and their errors have been calculated. In this study, factors affecting the trend such as the length of the time series, the variance of the noise and the autocorrelation of the noise are investigated. Special emphasis has been placed on the calculation of the statistical significance of the observed trends, which reveal significant local changes from $-5\%$ per year to $+5\%$ per year. These significant trends are distributed over the whole globe. Increasing trends have been calculated for Greenland, East Europe, Siberia and Oceania, whereas decreasing trends have been observed for the northwest USA, Central America, Amazonia, Central Africa and the Arabian Peninsular.

The idea of the comprehensive trend and significance analysis is to get evidence for the truth of these observed changes. While the significance estimation is based on intrinsic properties such as the length of the data sets, the noise and the autocorrelation, an important aspect of assessing the probability that the real trends have been observed is a validation with independent data.

Therefore an intercomparison of the global total column water vapour trends retrieved from GOME and SCIAMACHY with independent water vapour trends measured by radiosonde stations provided by the Deutsche Wetter Dienst DWD (German Weather Service) is presented.

The validation has been performed in a statistical way on the basis of univariate time series. Information about the probability of agreement between the two independently observed trends, conditional on the respective data, is revealed. On the one hand a standard t-test is used to compare the trends and on the other hand a

Bayesian model selection approach has been developed to derive the probability of agreement.

The hypothesis of equal trends from satellite and radiosonde water vapour data is preferred in 85% of compared pairs of trends. Substantial evidence for the hypothesis of agreeing trends is found in 26% of analysed trends. However, also disagreement has been observed, where the main reason has been identified on the one hand as the different spatial resolutions of the instruments. This means, that the radiosonde measurements can resolve very localised events, which is not possible with the satellite instruments. On the other hand, in contrast to the in principle continuously available (on a monthly mean basis) GOME/SCIAMACHY data, missing data in the radiosonde time series lead to trend discrepancies.

The identification and validation of water vapour trends is an important step for a better understanding of climate change, but water vapour is not the only contributing quantity. Beside water vapour, decisive parameters are temperature, clouds, precipitation, vegetation and many more. A promising framework for the investigation of a multivariate data set of environmental variables is given by the Markov chain analysis. As a first approach, the Markov chain analysis has been applied to a bivariate water vapour – temperature data set, where the global near surface temperatures are provided by the Goddard Institute of Space Studies (GISS) and cover a time span from 1880 to 2005. The temperature data are retrieved from ground stations and are mainly based on the Global Historical Climatology Network (GHCN).

In the framework of a Markov chain analysis, the bivariate set of data is reduced to a univariate sequence of symbols, which can be described as a discrete stochastic process, a Markov chain. This Markov chain represents the water vapour – temperature interaction or state of a region. Several descriptors have been calculated, such as *persistence, replacement of* and *entropy*. This approach is new in environmental science.

Exemplarily two climate systems, the Iberian Peninsular and a region at the islands of Hawaii in the central Pacific Ocean, are investigated. The Markov chain analysis is able to retrieve significant differences between the two climate systems in terms of the characteristic descriptors, which reflect properties such as climate stability, rate of changes and short term predictability.

# Publications

## Peer Reviewed Journal Articles

Mieruch, S., Noël, S., Bovensmann, H., and Burrows, J. P.: Analysis of global water vapour trends from satellite measurements in the visible spectral range, Atmos. Chem. Phys., 8, 491–504, 2008.

Noël, S., Mieruch, S., Bovensmann, H., and Burrows, J. P.: Preliminary results of GOME-2 water vapour retrievals and first application in polar regions, Atmos. Chem. Phys., 8, 1519–1529, 2008.

## Articles in Conference Proceedings

Mieruch, S., S. Noël, H. Bovensmann and J. P. Burrows, Verification of SCIAMACHY level 1 data by AMC-DOAS water vapour retrieval, Proc. 3rd Workshop on the Atmospheric Chemistry Validation of Envisat (ACVE-3), Frascati, Italy, 4-7 December 2006.

Noël, S., Mieruch, S., Buchwitz, M., Bovensmann, H., Burrows, J. P., 2006. GOME and SCIAMACHY global $H_2O$ columns. In: Proceedings of the First Atmospheric Science Conference. ESA Publications Devision, Noordwijk, The Netherlands.

Noël, S., Mieruch, S., Bovensmann, H., Burrows, J. P., 2007. A combined GOME and SCIAMACHY global $H_2O$ data set. In: ENVISAT Symposium 2007, SP_636_ENVISAT. ESA Publications Devision, Noordwijk, The Netherlands.

Melsheimer C., Mieruch S., Noël S., Heygster G., Comparison of Total Water Vapor Columns Retrieved from Satellite Measurements: Microwave Radiances from AMSU-B and Visible Spectra from GOME/SCIAMACHY, IEEE International Geoscience And Remote Sensing Symposium, Barcelona, July 2007.

# Awards

Mieruch S., Noël S., Reuter M., Bovensmann H., Schröder M., Schulz J., Burrows J.P., Global Water Vapor Trends From Satellite Data Compared With Radiosonde Measurements, AGU Chapman Conference on Atmospheric Water Vapor and its Role in Climate, Hawaii, October 2008, First price for the poster presentation in the climate session.

# Contents

1 Introduction    9

2 Fundamentals    13
    2.1 Earth's atmosphere . . . . . . . . . . . . . . . . . . . . . . . . . . 13
        2.1.1 Vertical structure of the atmosphere . . . . . . . . . . . . . . 14
        2.1.2 Greenhouse effect and climate change . . . . . . . . . . . . . 14
        2.1.3 Atmospheric water vapour and the hydrological cycle . . . . . 16
        2.1.4 The water molecule and water absorption . . . . . . . . . . . 18
    2.2 The GOME and SCIAMACHY instruments . . . . . . . . . . . . . . 19
        2.2.1 The GOME instrument on ERS-2 . . . . . . . . . . . . . . . . 19
        2.2.2 The SCIAMACHY instrument on ENVISAT . . . . . . . . . . 20
    2.3 Statistics . . . . . . . . . . . . . . . . . . . . . . . . . . . . . . . . 21
        2.3.1 Frequentist statistics vs. Bayesian statistics . . . . . . . . . . . 22
        2.3.2 Statistics in climatology . . . . . . . . . . . . . . . . . . . . . 24

3 The water vapour data set    25
    3.1 AMC-DOAS Retrieval . . . . . . . . . . . . . . . . . . . . . . . . . 25
        3.1.1 The AMC-DOAS principle . . . . . . . . . . . . . . . . . . . 25
        3.1.2 Present state of the AMC-DOAS product . . . . . . . . . . . . 26
    3.2 The combination of GOME and SCIAMACHY data . . . . . . . . . . 26
        3.2.1 Possible causes of the level shift . . . . . . . . . . . . . . . . . 27
        3.2.2 The seasonal component . . . . . . . . . . . . . . . . . . . . 29

4 Water vapour trends    31
    4.1 Trend estimation . . . . . . . . . . . . . . . . . . . . . . . . . . . . 31
        4.1.1 The trend model . . . . . . . . . . . . . . . . . . . . . . . . 31
        4.1.2 Global trend patterns . . . . . . . . . . . . . . . . . . . . . . 34
    4.2 Significance of trends . . . . . . . . . . . . . . . . . . . . . . . . . . 35
    4.3 Global trend . . . . . . . . . . . . . . . . . . . . . . . . . . . . . . 38
        4.3.1 Globally averaged water vapour trend . . . . . . . . . . . . . 39
        4.3.2 Influence of El Niño 1997/1998 on the global trend . . . . . . 40
        4.3.3 Water vapour correlation with temperature - Granger causality 42

| | | |
|---|---|---|
| **5** | **Comparison of water vapour trends** | **49** |
| | 5.1 Intercomparison of satellite and radiosonde trends | 49 |
| | 5.2 The radiosonde water vapour data | 50 |
| | 5.3 Regression analysis of satellite and radiosonde data | 51 |
| | 5.4 Student's t-test applied to trends | 52 |
| | 5.5 Bayesian model intercomparison | 53 |
| | 5.6 Analytical approximation | 57 |
| | 5.7 Application to water vapour | 59 |
| **6** | **Stochastic description of water vapour and temperature** | **67** |
| | 6.1 Interaction of water vapour and temperature | 67 |
| | 6.2 The Markov chain | 69 |
| | 6.3 Data sources | 70 |
| | 6.4 Preprocessing and construction of the Markov chains | 71 |
| | 6.5 Estimation of transition probabilities | 73 |
| | 6.6 Descriptors of the HTI | 76 |
| | 6.7 Significance of the descriptors | 78 |
| **7** | **Conclusions and outlook** | **83** |
| **A** | **Derivation of the error of a trend** | **89** |
| **B** | **Student's t-test** | **91** |
| **C** | **Trend estimation in matrix notation** | **93** |
| **D** | **Bayes' theorem** | **97** |
| **E** | **Bayesian model selection** | **101** |
| **F** | **Analytical approximation – the matrices** | **105** |
| **Bibliography** | | **107** |

# 1 Introduction

Water vapour is the most important natural greenhouse gas in the atmosphere and plays a crucial role in the context of climate change, because of strong feedback mechanisms (Held and Soden, 2000). Water vapour is a key player in atmospheric chemistry, e.g. the rapid conversion of sulfur trioxide to sulfuric acid, it is a source of the OH radical, and is also important for the ozone chemistry (Stenke and Grewe, 2005). Thus the knowledge of the global distribution of water vapour and its evolution in time is of utmost importance for climate system studies.

The strong infrared radiation absorbing character of water vapour generates the natural greenhouse effect. Without water vapour the global mean temperature at the surface would be 20 °C lower than today (Häckel, 1999). In this context the transport of water vapour constitutes an important aspect in the climate system. Atmospheric water vapour represents the movement of energy in the form of latent heat. By condensation this latent heat can be released yielding a warming of the atmosphere, which affects global circulation systems associated with weather and climate.

The Earth's surface temperature results from an equilibrium state of the incoming solar radiation and the outgoing terrestrial radiation. Changes in the atmospheric composition, especially those of greenhouse gases such as water vapour ($H_2O$), carbon dioxide ($CO_2$) and methane ($CH_4$) can alter the outgoing terrestrial radiation which leads to a new equilibrium state between the incoming and outgoing radiation fluxes, thus resulting in a changing Earth surface temperature. This has been reported by the Intergovernmental Panel on Climate Change (IPCC, 2007), which is a scientific intergovernmental body, commissioned to evaluate the risks of climate change. Carbon dioxide and methane, which are also measured with the SCIAMACHY instrument (Buchwitz et al., 2006; Schneising et al., 2008), are particularly important in the discussion of the anthropogenic greenhouse effect.

In the debates about climate change and the greenhouse effect, climate models predict e.g. a global increase of water vapour contents due to the global warming caused by increasing $CO_2$ and other greenhouse gases (Dai et al., 2001). This increased water vapour reduces the outgoing long-wave radiation, which yields to an additional warming of the troposphere (IPCC, 2007). Together with these indirect effects on the atmospheric water vapour contents, direct influences of anthropogenic interventions such as irrigation (Boucher et al., 2004) and deforestation (Gordon et al., 2005) alter the water vapour cycle and thereby the concentrations on local as well as on global scale. Therefore, a global monitoring of the atmospheric water vapour content is needed, which can be achieved using remote satellite sensing. The global water vapour total column amounts used in the present study have been retrieved by the Air Mass Corrected Differential Optical Absorption Spectroscopy approach (AMC-DOAS) (Noël et al., 2004) from spectral data measured by the Global Ozone Monitoring Experiment (GOME) flying on ERS-2 which was launched in April 1995 and the SCanning Imaging Absorption spectroMeter for Atmospheric CHartographY (SCIAMACHY) onboard ENVISAT launched in March 2002. The complete amount of water vapour is given in grams per atmospheric column on a $1\,\text{cm}^2$ base (unit: $g/cm^2$). For the trend study, the data set is spatially gridded on a global $0.5° \times 0.5°$ lattice and averaged over time on a monthly mean basis. Thus we are dealing with 259200 time series each with a length of 144 months (minus a few data gaps).

The water vapour column of the atmosphere can be seen as a proxy for the climate state of a region, whether it is, for instance, humid or dry. Moreover, it is strongly linked to the surface temperature of air. This strong correlation is shown by Wagner et al. (2006) for water vapour columns retrieved from GOME. The water vapour column amounts are high in the tropics, low over the poles and medium over the temperate zone. Figure 1.1 shows as an example the global annual mean of the water vapour column amounts for the year 2006 retrieved by the AMC-DOAS method (cf. Sect. 3.1.1) from SCIAMACHY data.

The water vapour trends can be seen as tracers following the climate state of a specific region. A decreasing trend, for example, could be a change from a humid state to a dry state of a specific region. An infinitely decreasing trend is impossible, so the trend has to stagnate at a certain point. If the water vapour columns have significantly changed, dramatic consequences for the flora (major vegetation types, savanna, tundra etc. as reported by Melillo (1999)), fauna and agriculture cannot be ruled out. Such changes would also affect and interfere with human society. Moreover this new state could be stable and a way back is perhaps not easy, or, connected with a strong hysteresis as shown by Scheffer and Carpenter (2003) in the framework of bifurcation analysis. The same arguments are valid for increasing trends vice versa.

The water vapour columns and their changes are strongly linked to the climate state and the vegetation type of a region. Plants, animals and humans are adapted to their environmental conditions. Changes or trends of the atmospheric water vapour columns, e.g. to dryer or more wet situations, can have critical consequences for life. Moreover, water vapour trend calculations are important to assess the quality of model results and increase our knowledge of the hydrological cycle on global and local scale.

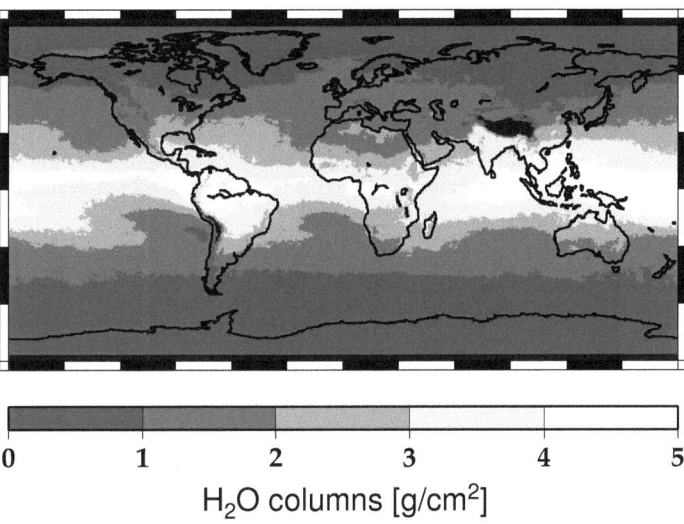

Figure 1.1: Annual mean of water vapour column amounts for the year 2006 derived from SCIAMACHY measurements. High water vapour columns are found near the equator, especially over rainforests. Small water vapour amounts are observed near the poles.

The water vapour trend study comprises the years 1996 to 2007, i.e. 12 years of global satellite data. This length of data cannot resolve long-term oscillation. However, it is enough to show significant water vapour changes in several regions on Earth.

This thesis is subdivided into 7 chapters. Chapter 1 is this introductory part. In Chap. 2 some fundamentals regarding the Earth atmosphere, water vapour and the greenhouse effect are discussed. Additionally, the role of statistics in environmental science is highlighted, which includes a short discourse through the field of standard mathematical statistics and Bayesian statistics. In Chap. 3 the water vapour retrieval method is explained schematically. Then, the combination of the GOME and SCIAMACHY data sets is presented. The water vapour trends are estimated in Chap. 4 including a significance analysis. Furthermore, the global averaged trends are investigated and correlated with global near surface temperature measurements. Chapter 5 deals with the validation of the satellite water vapour trends with independent water vapour trends from ground stations measured with radiosondes. Since water vapour is strongly correlated with the near surface temperature an analysis of the interaction of water vapour and temperature is presented in Chap. 6. To end this, the combined water vapour – temperature data set has been described as a stochastic process, a Markov chain. Chapter 7 gives the conclusions and an outlook. Finally, the appendix gives supplementary information about standard statistical methods and Bayesian methods used in this thesis.

# 2 Fundamentals

## 2.1 Earth's atmosphere

Life, as we know it today, would not exist without the Earth's atmosphere. The main meteorological and physical processes in the atmosphere are constituted by the solar radiation and its spatial and temporal variability due to the Earth's rotation (Roedel, 2000). The solar radiation interacts with the ocean, the land and the atmosphere and additionally, the gravity of the Earth contributes to the main forcing processes. The radiation budged is in principle given by:

- The incoming solar radiation with a maximum at about 500 nm. This radiation is absorbed and scattered in the atmosphere and at the Earth's surface.

- The thermic back reflexion from the Earth's surface and the atmosphere with a maximum at about $10\,\mu$m. This radiation is partly absorbed by the surface and the atmosphere and partly lost in space. Overall, there is a net balance between thermic infrared radiation and incoming solar radiation.

- Other heat and energy transport without radiation processes.

The chemical composition of the atmosphere is made of several gases with different properties. The following Tab. 2.1 shows the main constituents of dry air, which are in principle constant over spatial scales (up to about 90 km) and time scales of 10000s of years or even the age of the Earth. The composition of air regarding these gases is widely homogeneously distributed up to a height of about 80 km. Beside these permanent components, aerosols and several trace gases with life times from hours to years are present in the atmosphere. Thereunder we have the water vapour ($H_2O$), with 99.99 % in the troposphere and carbon dioxide ($CO_2$) with about 355 ppm in the atmosphere. Furthermore, without the claim of completeness, there are methane ($CH_4$) (ca. 1.7 ppm), hydrogen ($H_2$) (ca. 0.5 ppm), ozone ($O_3$) (ca. 0.05-5 ppm), nitrous oxide (laughing gas) ($N_2O$) (ca. 0.3 ppm), several nitrogen oxides (ca. 0.01-50 ppm), carbon monoxide (CO) (ca. 0.1 ppm), sulphur compounds (ca. 0.1-100 ppb (parts per billion)), ammonia ($NH_3$) (ca. 1-20 ppb) and many others.

| Constituent | Symbol | Volume percent |
|---|---|---|
| Nitrogen | $N_2$ | 78.09 |
| Oxygen | $O_2$ | 20.95 |
| Argon | Ar | 0.93 |
| Neon | Ne | $18.2 \cdot 10^{-4}$ |
| Helium | He | $5.24 \cdot 10^{-4}$ |
| Krypton | Kr | $1.14 \cdot 10^{-4}$ |
| Xenon | Xe | $0.087 \cdot 10^{-4}$ |

Table 2.1: Chemical composition of dry air from Roedel (2000). Note, that $10^{-4}$ volume percent corresponds to one part per million (ppm).

### 2.1.1 Vertical structure of the atmosphere

A common division of the atmosphere is based on the temperature profile. Figure 2.1 shows the US standard atmosphere defined 1976, which still constitutes a reference in several research activities. The mean temperature of air at the surface amounts to about 15 °C, which results mainly from the incoming solar radiation and the backscattered radiation from greenhouse gases (cf. Sect. 2.1.2). With increasing height, the temperature of air decreases until 10 to 13 km in high and tempered zones and until about 18 km in the tropics. This area is called troposphere and ends at a minimum temperature at about -50 °C to -55 °C in tempered zones and -80 °C in the tropics, called tropopause. Above this boundary, which is the stratosphere, the temperature increases again, which is caused by the absorption of ultraviolet radiation at wavelengths above 242 nm (Prölss, 2001) by ozone. The stratosphere ends at a maximum temperature at about 0 °C at around 50 km (Roedel, 2000), which is the stratopause.

Thereafter the temperature decreases (cf. Fig. 2.1) until the absolute minimum is reached at 80-90 km. This region is called mesosphere and the boundary at the minimum is the mesopause. Above this minimum the temperature again increases (which cannot be seen in Fig. 2.1) in this region, which is denoted as thermosphere, and converges above 200 km to about 1000 °C. The reason for the high temperature is the particles mean free path length of several kilometres, due to the low density of air at this height.

### 2.1.2 Greenhouse effect and climate change

The global mean temperature of the Earth amounts to approximately 288 K. According to the Stefan-Boltzmann law and assuming a thermic emissivity of 95 %,

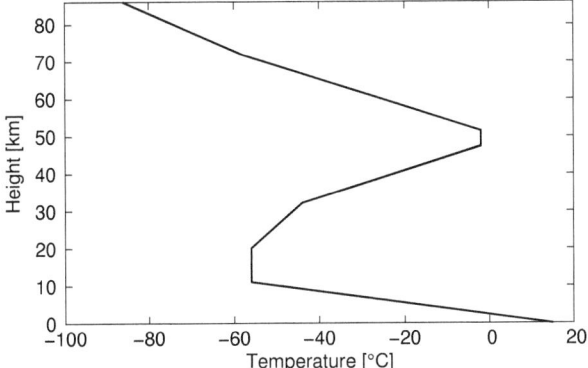

Figure 2.1: US standard atmosphere 1976.

the outgoing thermic infrared radiation would yield about 373 W/m². The total incoming solar radiation at the Earth surface is instead ca. 175 W/m² and at the top of the atmosphere about 342 W/m². This seems to be, on a first inspection, like a decisive discrepancy, because the equilibrium between incoming and outgoing radiation is violated. This alleged contradiction is solved by the fact, that the atmosphere absorbs infrared radiation and thus constitutes an emitter by itself. The backscattered radiation from the atmosphere to the Earth's surface amounts to ca. 300 W/m², thus the net outgoing radiation is about 73 W/m², which is below the incoming energy (Roedel, 2000). This shielding effect of the atmosphere is often compared with a glasshouse, which is transparent for the incoming short wave radiation and reserves the backscattered long wave radiation. This scenario is popularly named "greenhouse effect". Without the "greenhouse effect" the mean Earth's surface temperature would be $-15\,°C$ instead of the actual global mean temperature of $+15\,°C$. The main responsibility for the absorption of thermic radiation can be attributed to water vapour, $CO_2$, $CH_4$, $O_3$, $N_2O$ and also clouds. The natural amount of several greenhouse gases has increased decisively since the pre-industrial times, which is caused by human activities e.g. combustion of fossil fuels. Table 2.2 shows the increase of three major anthropogenically influenced greenhouse gases.

| Species | 2005 | 1750 | Increase [%] |
|---|---|---|---|
| $CO_2$ [ppm] | $379 \pm 0.65$ | $277 \pm 1.2$ | 37 |
| $CH_4$ [ppb] | $1774 \pm 1.8$ | $715 \pm 4$ | 148 |
| $N_2O$ [ppb] | $319 \pm 0.12$ | $270 \pm 7$ | 18 |

Table 2.2: Increase of long-lived greenhouse gases since the start of the industrial era (IPCC, 2007).

### 2.1.3 Atmospheric water vapour and the hydrological cycle

Water vapour is the gas phase of water and can be produced by evaporation of water and sublimation of ice. 99.99% of the atmospheric water vapour are located in the troposphere. The Earth's atmosphere contains about $13 \cdot 10^{15}$ kg or $13 \cdot 10^{12}$ m$^3$ water, which is mostly water vapour. Regarding the mean precipitation rate of about 1000 mm per year, the mean lifetime of water vapour amounts only to about 10 days (Roedel, 2000). This fast exchange is embedded into the global hydrological cycle, which describes the movement of water in the reservoirs, ocean, land, and atmosphere. The water cycle constitutes a closed system, thus the overall water content is constant over time. Figure 2.2 shows schematically the hydrological cycle.

The movement of water is initialised by the energy from the sun. Water evaporates from the oceans and fresh water reservoirs. Additionally water evaporates from plants, which is called evapotranspiration, and sublimates from ice and snow, which is the direct phase transition between the solid and gaseous phase. The water vapour is transported with warmer air up into the atmosphere and distributed globally by winds. The uprising air cools down and water condenses to cloud particles, which fall out as precipitation. Over several reservoirs (cf. Fig.2.2) the water cycle is closed and the evaporation/precipitation mechanism can continue.

Although the water cycle content is constant over time, the distribution of water within the reservoirs can vary. For instance, during colder times or ice ages, more water is stored in the ice and snow reservoirs, whereas in warmer times more water is stored in the oceans and atmosphere. The IPCC (2007) estimates an amplification of the water cycle in the 21$^{st}$ century. This means, that dry regions get dryer and humid region get more humid. Although the hydrological cycle moves immense masses of water, human activities do influence the cycle, even in such a big system. The impact of human influence comprises amongst others:

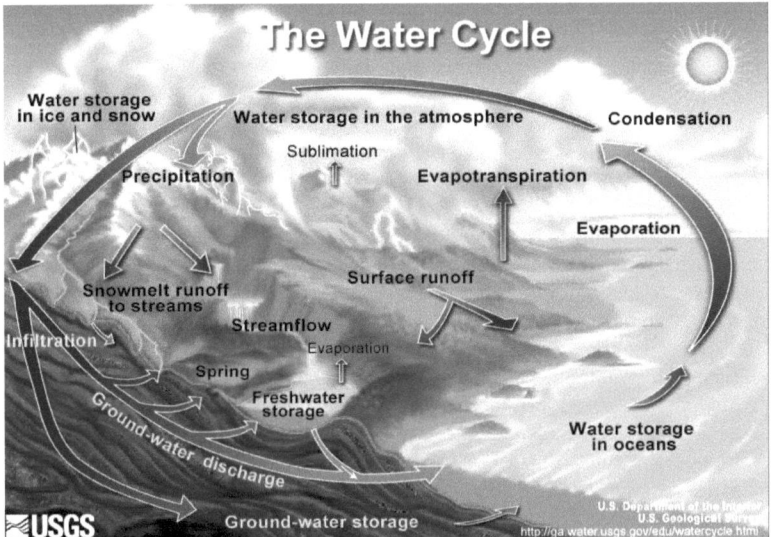

Figure 2.2: Global water cycle, graphic courtesy of the U.S. Department of the Interior U.S. Geological Survey.

- Agriculture.
- Alteration of the chemical composition of the atmosphere.
- Construction of dams.
- Deforestation and afforestation.
- Removal of groundwater from wells.
- Water abstraction from rivers.
- Urbanisation.

## 2.1.4 The water molecule and water absorption

The water molecule is the most abundant molecule on the Earth's surface. Water, at room temperature, is in principle colourless, tasteless, odourless and liquid. Water is an excellent solvent for many substances and it exists naturally in all three states of matter. Water is essential for life as we know it today.

Figure 2.3 depicts the structural formula of water. The molecule consists of two hydrogen atoms and one oxygen atom, which form a triangular. The oxygen atom resides at the vertex and the two hydrogen atoms span an angle of 104.45°. Since the electronegativity of oxygen is higher than that of the two hydrogens, the vertex of the molecule is negatively charged compared to the bottom side. Such difference in the electric charge is called a dipole. Due to the so-called dipole characteristic water can build hydrogen bonds with other water molecules. Additionally, the attraction of the oxygen and the two hydrogens is responsible for a strong bonding, which become apparent in the high boiling point (100 °C) and the high melting point (0 °C). These have to be seen in respect with chemically similar hydrogen compounds such as hydrogen sulfide ($H_2S$), hydrogen selenide ($H_2Se$) and hydrogen telluride ($H_2Te$), which have boiling points of -61 °C, -41 °C and -1 °C and melting points of -86 °C, -66 °C and -49 °C. Furthermore, the dipole is the reason, why water vapour absorbs thermic radiation strongly, because vibrational/rotational absorption only takes place, if there is a periodic change in the electric dipole moment of the molecule. Thus, water vapour strongly absorbs e.g. the backscattered thermic radiation of the Earth, from the microwave to the visible regions of the electromagnetic spectrum. This water vapour absorption is e.g. measured by SCIAMACHY. Figure 2.4 shows a typical measured spectrum from SCIAMACHY. The extraterrestrial solar irradiance is plotted in light grey and the backscattered radiance is shown in black. The main absorption bands from water vapour are clearly visible.

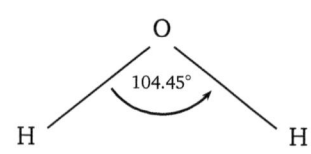

Figure 2.3: Structural formula of water vapour.

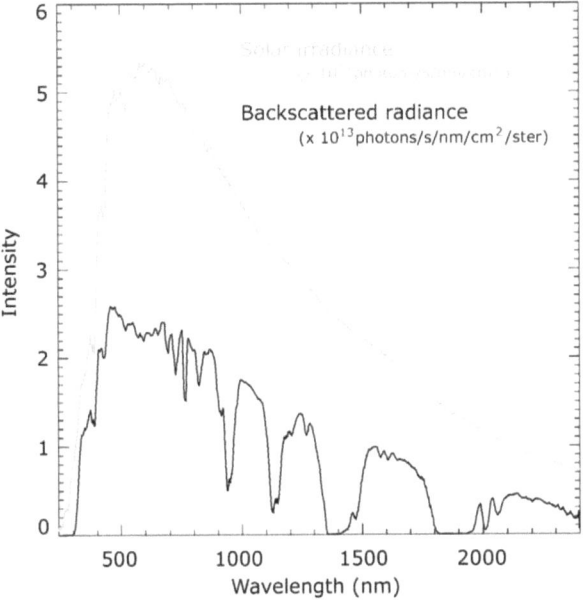

Figure 2.4: SCIAMACHY spectrum, before absorption takes place (grey) and afterwards (black). From Gottwald et al. (2006)

## 2.2 The GOME and SCIAMACHY instruments

### 2.2.1 The GOME instrument on ERS-2

GOME (Global Ozone Monitoring Experiment) is a passive imaging grating spectrometer on the European Remote Sensing satellite (launched on April 21$^{st}$, 1995), which flies on a sun synchronous orbit at an altitude of about 785 km. Therefore, a period of about 100 min is achieved, which is equivalent to about 14.3 orbits per day. ERS-2 has a global coverage of about three days and an equator crossing time at 10:30 local time. GOME measures the reflected, backscattered and transmitted solar radiation upwelling from the top of the atmosphere (Burrows et al., 1999) in nadir (downward) viewing mode. With a swath of about 960 km a resolution of ca. 40 km × 320 km is achieved. The spectral regions from 240 nm to 790 nm are captured and several trace gases such as ozone ($O_3$), nitrogen oxide ($NO_2$), oxygen ($O_2$), water vapour ($H_2O$), bromine monoxide (BrO), chlorine dioxide (OClO), sulphur dioxide ($SO_2$) and iodine monoxide (IO) can be measured.

### 2.2.2 The SCIAMACHY instrument on ENVISAT

SCIAMACHY (Greek: σκιαμαχια, "fighting shadows") (SCanning Imaging Absorption spectroMeter for Atmospheric CHartographY) is a passive imaging grating spectrometer on the Environmental Satellite ENVISAT (launched on March $1^{st}$, 2002). Similar to ERS-2, ENVISAT flies on a sun synchronous orbit at a height of about 785 km. It has a period of about 100 min with 14.3 orbits per day. ENVISAT crosses the equator at 10:00 local time. Thus GOME and SCIAMACHY cross each point on Earth with a time lag of 30 minutes. SCIAMACHY measures the reflected, backscattered and transmitted solar radiation upwelling from the top of the atmosphere (Burrows et al., 1990, 1995; Bovensmann et al., 1999; Gottwald et al., 2006). SCIAMACHY captures the spectral regions from 214 nm to 1773 nm continuously in six channels. Additionally, two channels from 1934 nm to 2044 nm and from 2259 nm to 2386 nm give information on infrared absorbing species. Amongst others, information on atmospheric gases and trace gases such as $O_3$, $NO_2$, $O_2$, $H_2O$, $CH_4$, $CO_2$, CO, BrO, OClO, $SO_2$ and IO can be retrieved with SCIAMACHY. Furthermore, information can be derived about aerosols and clouds. SCIAMACHY operates in three different measuring geometries, which are also shown in Fig. 2.5:

- Nadir-view: The nadir measuring mode captures the concentrations of several trace gases in the total atmospheric column on a 960 km wide swath orthogonal to the flight direction. The resolution is about 30 km in flight direction and about 60 km orthogonal to the flight direction.

- Limb-view: The limb mode allows to retrieve information on the vertical distribution of trace gases. The field of view in flight direction is about 2.6 km in the distance of about 3000 km. Orthogonal to the flight direction the resolution of the measurements accounts to 240 km. Thus, in limb mode the vertical atmosphere is sampled in 3 km steps.

- Occultation: In the occultation mode SCIAMACHY directly observes the sun or the moon through the atmosphere. The resolution is 30 km horizontal and 2.5 km vertical. An advantage of the occultation mode is the high precision of the measurements, but a disadvantage is the bad spatial coverage, because occultation is only possible during sun/moon rises (from the instruments view).

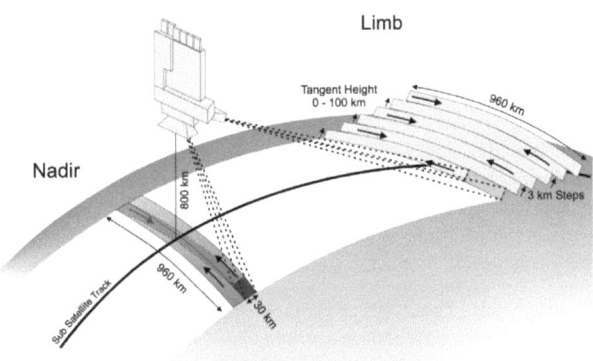

Figure 2.5: SCIAMACHY spectrometer in the field. The two measuring modes are shown, nadir and limb. Here it has to be noted, that the nadir and limb measurements are performed alternately and not simultaneously as insinuated by this viewgraph. Figure provided by S. Noël, IUP.

In this thesis, the water vapour data from GOME and SCIAMACHY are used, which have been retrieved from nadir measurements.

## 2.3 Statistics

Statistics is a mathematical science, which is embedded into the theory of probability. Its main objectives are the analysis, interpretation or explanation and presentation of data. It is widely used in natural science, social science, humanities, government and business and provides also methods for prediction and forecasting based on data.

The three major objectives of statistics are:

- **Description:** The descriptive statistics describe data, which have been recorded, with characteristic quantities such as the mean and the variance. Furthermore, the graphical presentation using diverse diagrams and histograms belongs to the description.

- **Exploration:** The data exploration goes a step further and can be summarised as the search for structures or characteristics in the data. Thus, a scatter-plot between two random variables can give e.g. information of possible correlations.

- **Induction:** The inductive statistics make massive use of probability theory and stochastics to find evidence for underlying processes or basic populations. The induction provides methods to answer questions such as: Are the data normally distributed? Is the observed trend statistically significant? Which theory is superior, $A$ or $B$?

Two major schools of statistics coexist, the frequentist statistics and the Bayesian statistics. The two different concepts are outlined in the next section.

### 2.3.1 Frequentist statistics vs. Bayesian statistics

The frequentist statistics approach was mainly developed by e.g. Fisher, Neyman and Pearson at the beginning of the 20$^{th}$ century. The underlying philosophy of the frequentist statistics is the interpretation of an events probability as the limit of its relative frequency in a large number of trials. A major component of how statistics are used in environmental science is the hypotheses testing, which is used under the framework of induction to make decisions using experimental data. The basic concept of hypothesis testing is to set up a null-hypothesis $H_0$, which is assumed to be true and an alternative hypothesis $H_1$, which is the complementary event of $H_0$. Then the probability of observing a value of a test statistic (according to $H_0$) that is at least as extreme as the value that was actually observed is inferred. The null-hypothesis is typically rejected, if the observed probability is below 0.05, which is also called the 95% confidence level. Such a case would confirm the alternative hypothesis.

The Bayesian statistics have been developed by Bayes (1763) and de Laplace (1812), thus they are much older than the frequentist approach, but were then largely forgotten, until Jeffreys (1939) rediscovered the ideas of Bayes and de Laplace. The Bayesian concepts have undergone a renaissance in the late 20$^{th}$ century, amongst others by the increase of computational power. Important impact on the Bayesian development in recent times have e.g. Jaynes and Bretthorst (2003). An advantage of the Bayesian formalism is, that it is based completely on probability theory, whereas the frequentist statistics have not such an basic underlying concept, and are rather a compilation of a large amount of diverse tests and methods. As in the frequentists approach, the central point of hypothesis testing can also be accomplished within the Bayesian framework. However, Bayesian hypothesis testing can better be described as a model selection procedure, i.e. infering, which model or hypothesis has the higher probability in explaining certain data or phenomena. The mathematical derivation of the model selection method is given in the App. E.

The two major differences between frequentist statistics and Bayesian methods are:

- **Philosophical difference**: The deep philosophical difference between Bayesian theory and frequentist statistics is, that the Bayesians draw conclusions about the relative evidence for parameter values given a data set, while frequentists estimate the relative chance of data sets given a parameter value. This can be elucidated in the sense of conditional probabilities. A conditional probability is the probability of a proposition $X$ given the occurrence of another proposition $Y$ and is denoted as $P(X|Y)$. In the frequentist approach $X$ could be e.g. a parameter value and $Y$ could be the null-hypothesis of some underlying properties e.g. normal distribution. Following, one would derive the probability of the parameter assuming that the null-hypothesis is true. The Bayesian concept can give the reverse, i.e. the probability of a hypothesis if a certain parameter has been observed $P(Y|X)$. Hence, generally the frequentist statistics can infer the probability of data or parameters given the null-hypothesis $P(data|hypothesis)$, while Bayes' theorem can give the reverse, the probability of the hypothesis given the data or parameters, which is $P(hypothesis|data)$.

- **Prior information**: Bayesian methods comprise prior information about the truth of a hypothesis or parameter range, which reflects the knowledge (or ignorance) before the data have been analysed. In frequentist statistics such prior information does not exist.

The fundament of Bayesian statistics is given by Bayes' theorem, which can be formulated as:

$$P(Y|X,I) = \frac{P(X|Y,I) \cdot P(Y|I)}{P(X|I)}, \qquad (2.1)$$

where $X$ and $Y$ are propositions and $I$ denotes relevant background information. The $I$ is often neglected, but it has to kept in mind, that no absolute probabilities exist without certain background assumptions or information. $P(Y|X,I)$ is called the posterior probability, $P(X|Y,I)$ is the likelihood, $P(Y|I)$ is the prior probability and $P(X|I)$ has former been called the marginalization likelihood, but nowadays Sivia and Skilling (2006) introduced the term 'evidence' for the denominator (more information is given in the appendix).

### 2.3.2 Statistics in climatology

Climatology is, in a large part, the study of the statistics of our climate (Storch and Zwiers, 1999). Mathematical statistics are widely used from simple methods, such as the mean and variance, to sophisticated concepts, which reveal the dynamics of the climate system.

Our climate is a nonlinear dynamical system, which is mainly driven by large external forcing like the solar radiation. But the climate is also influenced by seemingly marginal phenomena like flapping butterflies (Storch and Zwiers, 1999). This is also founded in the works of Lorenz (1963), who has built the fundament of the theory of chaotic systems.

Although the climate system is generally a deterministic system we cannot describe it deterministically, because we do not know all factors controlling the climate mechanisms. Therefore we use probabilistic concepts and statistical methods to describe the climate. A hitherto successful (and often the only possible) strategy is to analyse only a few number of climate parameters and identify the rest as background noise. Often the noise is interpreted as nuisance, but it can also be seen as an important information of the system. Furthermore, nonlinearities and instabilities are responsible for the unpredictability of the climate beyond certain times, which is an argument for the use of probabilistic approaches.

# 3 The water vapour data set

## 3.1 AMC-DOAS Retrieval

### 3.1.1 The AMC-DOAS principle

The global water vapour total column amounts used in the present study have been retrieved by the Air Mass Corrected Differential Optical Absorption Spectroscopy approach (AMC-DOAS) (Noël et al., 2004) from spectral data measured by the Global Ozone Monitoring Experiment (GOME) flying on ERS-2 which was launched in April 1995 and the SCanning Imaging Absorption spectroMeter for Atmospheric CHartographY (SCIAMACHY) onboard ENVISAT launched in March 2002. The basic principle of the method is to calculate the difference between the measured Earthshine radiance and the solar irradiance at wavelengths where water vapour absorbs radiation (here the wavelength band from 688 nm to 700 nm is used) and relate this absorption-depth to the water vapour column concentration. Because visible measurements are restricted to daylight conditions and almost cloud free scenes, the AMC-DOAS method provides in principle a cloud free daytime water vapour climatology, however it can also be applied to partially cloudy scenes. This is achieved using an air mass correction factor (AMCF) based on the $O_2$ column (Noël et al., 2004). Within the AMC-DOAS retrieval certain surface and atmospheric conditions are assumed, namely no surface elevation, a surface albedo of 0.05, a tropical atmosphere and especially the absence of clouds. Usually these conditions differ from the real ones, which is accounted for by the AMCF derived from $O_2$ absorption. Via the AMCF the water vapour columns are scaled such that the correct $O_2$ optical depth is achieved (see Noël et al. (2004) for details). Deviations of the AMCF from unity indicate discrepancies between the assumed and the real conditions and if these deviations are too large (AMCF < 0.8), the water vapour measurements are discarded. One of the main reasons for AMCF's differing from unity is the presence of clouds in the observed scene. Therefore the AMCF limit efficiently sorts out too cloudy scenes, but it is possible to derive water vapour columns also from partly cloudy scenes, as long as the cloud fraction is low (AMCF ≥ 0.8). In this sense the AMC-DOAS products provide a cloud-cleared climatology.

The AMC-DOAS method most probably slightly underestimates the water vapour columns in cloudy cases, because contrariwise to the well mixed $O_2$, the water vapour volume mixing ratio increases towards the surface. Thus the AMCF for water vapour should be typically lower than that for $O_2$. However, this second order effect affects both, the GOME and SCIAMACHY measurements in the same way and should not influence the water vapour trends in contrast to trends in the cloud cover which can most probably influence the water vapour trends. A qualitative estimation of the impact of clouds on the water vapour trends would be on the one hand the observation of lower trends, if an increase in the cloud cover over time would occur, because of increasing underestimated measurements with time. A positive biased trend could be caused by a decrease in the cloud cover over time, because then less data with a negative bias are measured. It has to be noted, that this is also a second order effect, because the climatology is in principle cloud free (only data with AMCF $\geq$ 0.8), but it cannot be excluded.

### 3.1.2 Present state of the AMC-DOAS product

The AMC-DOAS method provides a completely independent data set, because it does not rely on any additional external information. The retrieval of water vapour data from the GOME instrument is described in Noël et al. (1999), where also validation results of the data with SSM/I (Special Sensor Microwave Imager) data are shown. Likewise, SCIAMACHY water vapour data have been validated with SSM/I and ECMWF (European Centre for Medium-Range Weather Forecasts) data (Noël et al., 2005). An intercomparison and a preliminary connection of both, the GOME and the SCIAMACHY data sets, is shown in Noël et al. (2006). The high quality of the two water vapour data sets is demonstrated from validation and comparison results, which shows that they can be merged well together. Thus, the trend analysis presented in this thesis is build on a solid fundament. Furthermore, the water vapour data are gridded on a 0.5° × 0.5° lattice and averaged to monthly means, which are representative for the respective months. A good overview of other water vapour measuring instruments from space can be found in Brocard (2006). Previous investigations of other water vapour retrievals from GOME are described e.g. in Maurellis et al. (2000) and Lang et al. (2003). A similar water vapour trend study to this is presented by Wagner et al. (2006) for the GOME data, based on a different retrieval method described in Wagner et al. (2003).

## 3.2 The combination of GOME and SCIAMACHY data

GOME on ERS-2 has been measuring since June 1995 up to the present, but since June 2003 no global coverage is provided as a result of a breakdown of the on-board tape recorder. SCIAMACHY data are available since August 2002, but the

SCIAMACHY instrument did not achieve final flight conditions until January 2003. The quality of the SCIAMACHY water vapour data is furthermore slightly reduced in 2002, because of the non-availability of an actual solar reference spectrum prior to December 2002. Overall the most appropriate time for the change from GOME to SCIAMACHY data results in January 2003.

When combining the data sets possible level shifts between GOME and SCIAMACHY measurements have to be accounted for. Therefore the period of near simultaneous global measurements of GOME and SCIAMACHY, August 2002 to June 2003, has been studied explicitly. The global agreement results in an average deviation of $-0.01\,\mathrm{g/cm^2}$ with a scatter of $\pm 0.25\,\mathrm{g/cm^2}$ (Noël et al., 2007). This means, that on a global mean, there is strictly speaking no difference between the results of both instruments. This is anticipated, because the same retrieval method (AMC-DOAS) is used for both instruments and the method is quite insensitive to existing calibration differences between the GOME and SCIAMACHY instruments. The scatter of the water vapour differences between the two instruments results from local (single grid pixel) time series, which show deviations. Although these differences on a local scale are small ($\pm 0.25\,\mathrm{g/cm^2}$) compared to the total water vapour column, they can influence the trend and have to be considered.

### 3.2.1 Possible causes of the level shift

The calibration between the instruments as a cause for the level shifts on local scale can be ruled out, because the AMC-DOAS method is quite insensitive to absolute radiometric calibration. Therefore two main aspects are considered to be responsible for the differences:

1. Different equator crossing time.
   GOME on ERS-2 and SCIAMACHY onboard ENVISAT, respectively, cross the equator at 10:30 and 10:00 local time. That means SCIAMACHY and GOME measure at different times slightly different states of atmospheric composition. It is most probable, that fluctuations in the water vapour column on fast time scales caused by e.g. winds and clouds are responsible for the level shifts. It follows that a possible mean level shift between both data sets has to be allowed for the combination of the data on a local scale.

2. Differing spatial resolutions.
   The spatial resolution of the GOME data is $40\,\mathrm{km} \times 320\,\mathrm{km}$, whereas it is (typically) $30\,\mathrm{km} \times 60\,\mathrm{km}$ for SCIAMACHY data. When combining both data sets, different (higher) seasonal amplitudes have been accounted in Mieruch et al. (2008) for the SCIAMACHY data with respect to GOME. Because of the

higher resolution of SCIAMACHY, higher peaks (negative as well as positive) of water vapour can be detected. However, it turned out, that the negligence of the possible amplitude change has only very marginal influence on the trend results. As described in Sect. 3.2.2 an amplitude change is not considered and the seasonal component is removed by calculating anomalies. But, the resolution together with the cloud cover contributes to the level shift. Due to the higher resolution, SCIAMACHY "sees" more cloud free pixels than GOME which introduces a potentially positive bias for the SCIAMACHY data. However, this bias is observed on local scale, it is not visible on average. As mentioned in Sect. 3.1.1 we expect a negative bias for the AMC-DOAS data, due to remaining clouds. Because of the different spatial resolutions, partly cloudy scenes are more probable for GOME; therefore a more negative bias for the GOME data compared to the SCIAMACHY data is expected. This is in line with the findings of higher SCIAMACHY columns and thus positive level shifts around the equator regions, where high cloudiness is more probable. Figure 3.1 shows the distribution of global level shifts, which have been estimated in a least square sense in Sect. 4.1.1.

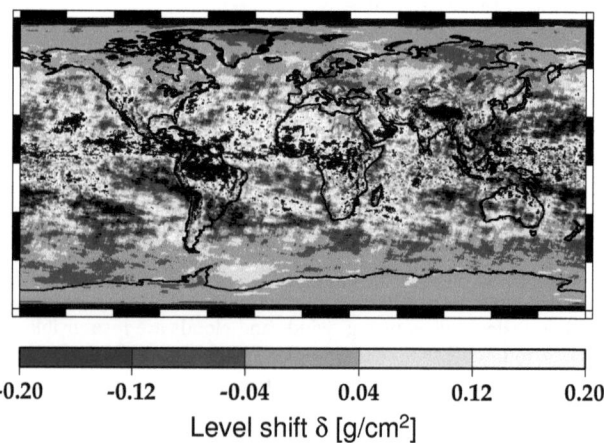

Figure 3.1: Level shift $\delta$ between GOME and SCIAMACHY measurements observed from the least square regression in Sect. 4.1.1.

The local level shifts result from a complex interaction of atmospheric processes (clouds, winds, small scale fluctuations, diurnal cycle) within the 30 minutes time delay of GOME and SCIAMACHY and instrumental differences (resolution).

The water vapour columns are retrieved on a daily basis, but it has to be noted that ERS-2 and ENVISAT fly on a sun-fixed orbit, i.e. passing each point on Earth at constant local time. Thus measurements from GOME and SCIAMACHY are snapshots of the actual atmospheric conditions at specific locations at specific times.

A global coverage is achieved for GOME data within 3 days and for SCIAMACHY nadir measurements within 6 days. Thus, in principle monthly mean data provide a data set without gaps. However, few gaps are observed even in the monthly mean data, because high cloudiness and high mountain area (e.g. the Himalayas) measurements are removed from the data by the AMC-DOAS algorithm. Moreover, since GOME and SCIAMACHY are spectrometers using the sunlight, measurements are only possible during daylight, and therefore no data is available at night, which results in a lack of measurements at the north pole and Antarctica during the polar nights. Since GOME and SCIAMACHY are measuring in the nadir viewing geometry no profile information of water vapour can be retrieved in this mode.

The derivation of water vapour columns from GOME-type instruments has also some unique advantages: The retrieval is possible over land and ocean and no external calibration sources like radiosondes are required. Although the resulting water vapour time series is quite short compared to other instruments like SSM/I which are looking forward to a 40 years series, it will be extended by other SCIAMACHY measurements and especially by the series of GOME-2 instruments on MetOp, of which the first one was launched successfully in 2006 (Noël et al., 2008). The series of GOME-type instruments has therefore the potential to provide independent and consistent water vapour data sets on both land and ocean for at least 25 years.

### 3.2.2 The seasonal component

The strong seasonal component, which is enclosed in the water vapour data because of the relation to temperature, can clearly be seen in the AMC-DOAS product. In Mieruch et al. (2008) the oscillatory parts have been described by a Fourier series

$$S_t = \eta \sum_{j=1}^{4} \left[ \beta_{1,j} \cdot \sin(2\pi j t/12) + \beta_{2,j} \cdot \cos(2\pi j t/12) \right], \tag{3.1}$$

on a monthly mean temporal grid, where $t$ denotes the time. The $\beta_{1,j}$ and $\beta_{2,j}$ are the Fourier coefficients which have been estimated in a least square sense.

$\eta = 1 + (\gamma - 1)U_t$ describes an amplitude change of magnitude $\gamma$ at time $t \geq T_0$, where $U_t$ is a step function:

$$U_t = \begin{cases} 0, & t < T_0 \\ 1, & t \geq T_0 \end{cases}, \qquad (3.2)$$

and $T_0$ is the point in time, when the GOME and SCIAMACHY data are merged together.

Another possibility to describe the harmonic components is the calculation of the seasonal cycle averaged over all years:

$$S'_n = \frac{1}{12} \sum_{i=0}^{T/12-1} Y_{i \cdot 12 + n} \qquad n = 1, ..., T/12 \qquad (3.3)$$

where $Y$ represents the monthly mean water vapour columns and $T$ is the total number of months. The mean seasonal component is quite equal to the Fourier description. It has to be noted, that trends or level shifts in the data are not influenced by calculating anomalies. The trend analysis has been performed for the two different approaches of modelling the seasonal terms and it turns out, that the trends and also the errors are quite independent of the choice of the above procedures. Therefore the averaged seasonal cycle has been chosen to deseasonalise the data, by means of less computational costs. It has to be noted, that the overall mean has been added to the anomalies to get the deseasonalised data.

# 4 Water vapour trends

## 4.1 Trend estimation

### 4.1.1 The trend model

The detection of trends is difficult and depends on the length of the time series, the magnitude of variability and autocorrelation of the data (Weatherhead et al., 1998). The trends can be influenced by level shifts inside the time series from instrument changes or new instrumental calibration etc.. Short time series as well as high variability, autocorrelation and level shifts in the data increase the uncertainty of trend detection. Statistical methods are used to reveal trends and explore their uncertainties. As discussed in Sect. 3.2 the analysis of overlapping GOME and SCIAMACHY data strongly supports the use of a level shift model. The methods used here are based on the approach of Weatherhead et al. (1998) and Tiao et al. (1990) and have been adapted to our needs. The time series of the data (with removed seasonal component) at one geolocation (i.e. a single grid point) can be described by the following trend model:

$$Y_t = \mu C_t + \omega X_t + \delta U_t + N_t, \qquad t = 0,...,T, \tag{4.1}$$

where $Y_t$ contains the water vapour measurements. $\mu$ is the mean water vapour column of the time series at time $t = 0$. $C_t$ is a constant, which is unity for all $t$ and needed for the following consideration of autocorrelations. $\omega$ represents the trend and $X_t$ contains the time. In the case of monthly averaged data the time span from January 1996 until December 2007 or from month 0 to 143 is considered. The data have not to be necessarily equidistant as there may be missing data. $\delta$ is the magnitude of a mean level shift at time $t = T_0 \, (0 < T_0 < T)$, where $T_0 = 84$ represents the intersection of GOME and SCIAMACHY data on January 2003. $U_t$ describes the former introduced step function Eq. 3.2.

The last term $N_t$ in Eq. (4.1) contains the unexplained portion of the data, i.e. the noise. The noise $N_t$ is assumed to be an autoregressive process of the order of one [AR(1)] (Schlittgen and Streitberg, 1997), i.e.

$$N_t = \phi N_{t-1} + \epsilon_t, \tag{4.2}$$

where $\epsilon_t$ are independent random variables with zero-mean and variance $\sigma_\epsilon^2$. This assumption is used because environmental data are often autocorrelated, e.g. if the temperature is high at one day, a high temperature is likely on the next day. The magnitude or the memory of the autocorrelation is presented by $\phi$, which is restricted to $-1 < \phi < 1$, so the noise process $N_t$ is stationary. The memory of the data at lag one can be calculated using the autocorrelation function $\phi = \text{Corr}_{N_t N_{t-1}}$, which is directly linked to the well known correlation coefficient.

Generally the autocorrelation function is restricted to continuous, statistically stationary stochastic functions, or in the discrete case equidistantly sampled data. Since there are gaps in our time series the discrete correlation function for analysing unevenly sampled data which was originally developed by Edelson and Krolik (1988) for astronomical problems, was applied.

To calculate the autocorrelation of the noise, the noise itself has to be determined by applying the model (Eq. (4.1)) to the data in a least square sense and subtract the fit from the data. The noise $N_t$ is then given by the remaining residuals:

$$N_t = Y_t - (\hat{\mu}C_t + \hat{\omega}X_t + \hat{\delta}U_t), \tag{4.3}$$

where $\hat{\mu}, \hat{\omega}, \hat{\delta}$ are the least square estimators. The $N_t$ are used to calculate first the set of unbinned discrete correlations

$$\theta_t = \frac{N_t \cdot N_{t-1}}{\sigma_N^2}, \qquad t = 1,...,T, \tag{4.4}$$

where the $N_t$ have zero-mean and variance $\sigma_N^2$. Following, the $\theta_t$ have to be assigned to their lags $\tau_t$ with

$$\tau_t = X_t - X_{t-1}, \qquad t = 1,...,T. \tag{4.5}$$

Now, the magnitude $\phi$ of autocorrelation at lag $\tau = 1$ can be determined by averaging over the number $M$ of $\theta_t$ with corresponding $\tau_t = 1$:

$$\phi = \frac{1}{M} \sum_{i=1}^{M} \theta_i(\tau_i = 1). \tag{4.6}$$

The mean autocorrelation function (from $\approx 259200$ time series) $\text{Corr}_{N_t N_{t-\tau}}$ of water vapour noise $N_t$ for lags $\tau$ from one to six months is shown in Fig. 4.1 as a black line together with the standard deviation (grey). As can be seen, the autocorrelation function is fast decreasing, thus the consideration of autocorrelation at lag one is quite convincing.

The aim of the above calculations concerning autocorrelations is to account for them during the fitting procedure. This is performed by a linear matrix transformation. Making the connection to the autoregressive process of Eq. (4.2), the model has absorbed the autocorrelations of $N_t$ into the transformed data $Y_t^*$, $C_t^*$ (which is

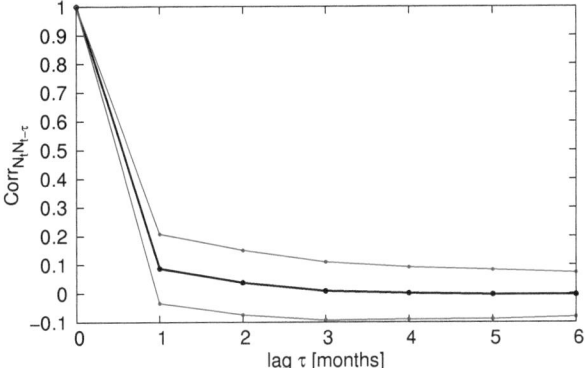

Figure 4.1: Mean autocorrelation function (black) of water vapour noise $N_t$, together with the standard deviation (grey).

no more constant), the time $X_t^*$ and the step function $U_t^*$, whereas the $N_t$ have lost their autocorrelations and have become white noise $\epsilon_t$:

$$Y_t^* = \mu C_t^* + \omega X_t^* + \delta U_t^* + \epsilon_t, \qquad t = 0,...,T. \tag{4.7}$$

Now a linear regression is applied, which can be solved analytically for the least square estimators $\hat{\mu}, \hat{\omega}, \hat{\delta}$ and their errors $\sigma_{\hat{\mu}}, \sigma_{\hat{\omega}}, \sigma_{\hat{\delta}}$. Details of the transformation and regression are given in the appendix and in Weatherhead et al. (1998).

After the implementation of the autocorrelations into the model and solving the linear least square equations (where the least square estimator of the trend is denoted with $\hat{\omega}$) Weatherhead et al. (1998) derive an approximation of the error of the trend $\sigma_{\hat{\omega}}$:

$$\sigma_{\hat{\omega}} \approx \frac{\sqrt{12}\,\sigma_N}{\ell^{\frac{3}{2}}} \cdot \sqrt{\frac{1+\phi}{1-\phi}} \cdot \frac{1}{[1-3\vartheta(1-\vartheta)]^{\frac{1}{2}}}. \tag{4.8}$$

$\sigma_{\hat{\omega}}$ depends on the standard deviation $\sigma_N$ of the noise, the length of the time series $\ell$, the autocorrelation $\phi$ of $N_t$ and the fraction $\vartheta = T_0/T$, which describes the position of the level shift. This approximation of the error has also been applied and compared with the non-approximated errors (Eq. C.11), which is shown as a

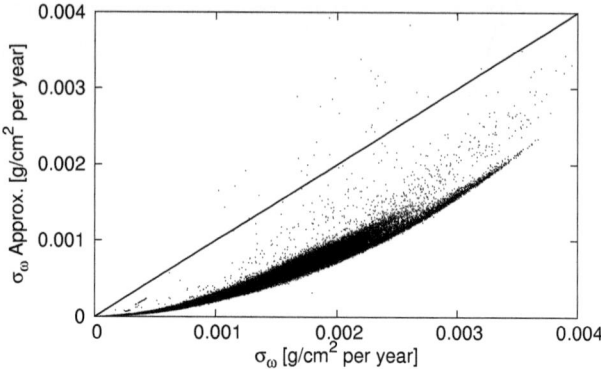

Figure 4.2: The errors of the trends using the assumption of Eq. 4.8 are plotted against the non-approximated errors derived in the appendix (Eq. C.11).

scatter-plot in Fig. 4.2. The approximated trend error in Fig. 4.2 is underestimating the non-approximated errors and the mean relative difference is of the order of 80%, thus it is not recommended to use the approximation for the water vapour data. Therefore, the errors of the trends have been calculated using Eq. C.11. However, the general dependencies of Eq. 4.8 are still valid.

### 4.1.2 Global trend patterns

The global trend patterns are determined from the long-term time series from January 1996 to December 2007 including GOME and SCIAMACHY globally gridded monthly mean data on a 0.5° × 0.5° grid. Two ways of investigating the trends are informative; on the one hand displaying the absolute trends $\hat{\omega}$ in g/cm² per year (Fig. 4.3) and on the other hand displaying the relative trends $\hat{\omega}/\hat{\mu}$ in % per year (Fig. 4.4), where $\hat{\mu}$ represents the deseasonalised water vapour columns at the beginning of the time series. The absolute trends shown in Fig. 4.3 are stronger near the equator and smaller near the poles. Dark as well as light patches are seen, thus there are negative as well as positive trends observed, however most trends are small and distributed around zero. For the relative trends the situations is inverted and we find larger relative trends at the poles than at the equator, because

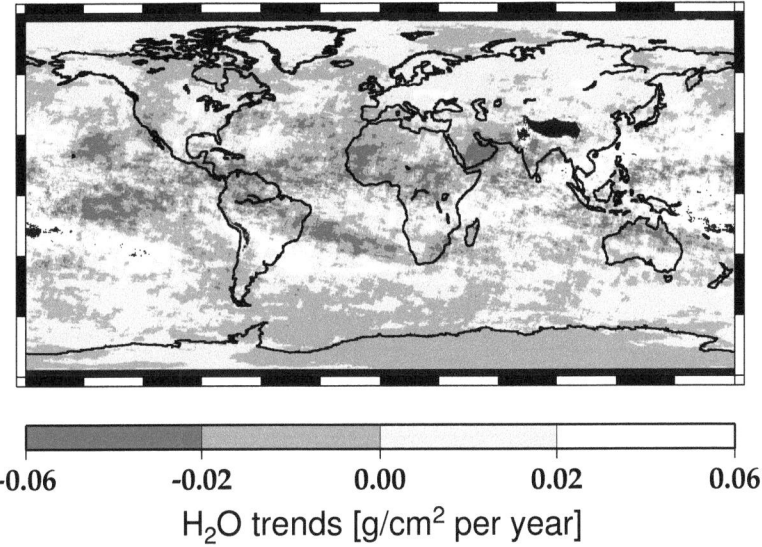

-0.06    -0.02    0.00    0.02    0.06
H$_2$O trends [g/cm$^2$ per year]

Figure 4.3: Global absolute water vapour trends.

the relative trends are normalised to the respective columns, which are small at the poles and large at the equator. High mountain areas, like the Himalayas and parts of the Andes are excluded from the trend calculation (indicated with black colour), because the water vapour retrieval is not possible for these extreme elevations.

## 4.2 Significance of trends

One main question concerning trends is whether the trend is significant or not. The answer to this question can only be given in a probabilistic sense. Here, the frequentists strategy is followed to estimate the significance of the trends.

Based on the null-hypothesis that the observed trend is equal to zero $H_0 : \hat{\omega} = 0$ the alternative hypothesis is the observation of a non-zero trend $H_1 : \hat{\omega} \neq 0$. The least square method assumes Gaussian distributed data around the fitted function. Using standard rules of random variables it can be shown that the trend $\omega$ is a linear function of the data $Y_t$ (cf. App. A) and therefore also Gaussian distributed (Fahrmeir et al., 2004). If it would turn out, that the probability of the data given

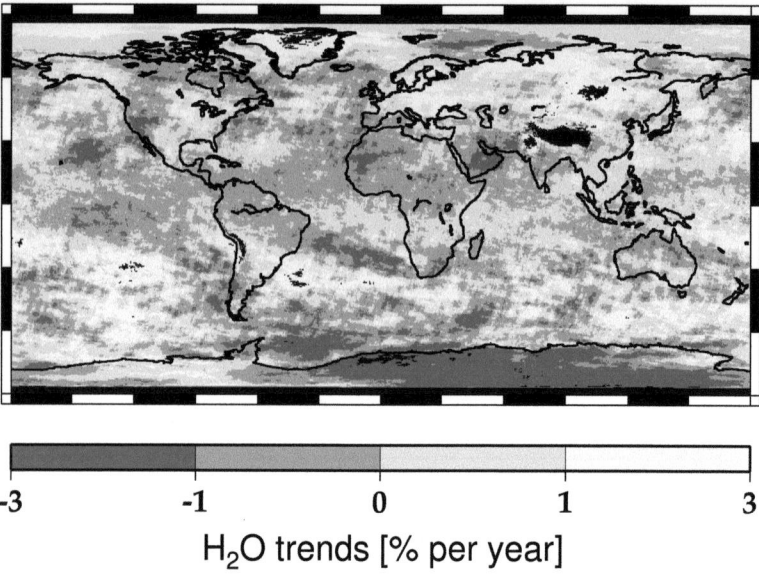

Figure 4.4: Global relative water vapour trends.

the nullhypothesis $P(D|H_0)$ is < 0.05, than the chance of making an error in rejecting the null hypothesis is 5%. Accordingly, the likelihood to be correct in confiding the alternative hypothesis is 95%, which does not mean, that the hypothesis is true with 95%. Actually a Student's t-test has to be used, because the error of the trend is not known and has to be estimated. However if the parameters are estimated from populations of more than 30 data points the t-test can be substituted by the Gauss-test (Fahrmeir et al., 2004), which is done here. The difference between the Gauss-test and t-test is simply, that the probability of the test statistic (which is the same for both methods) has to be looked up in the Gauss distribution or in the Student distribution, respectively. Figure 4.5 shows the results of the significance analysis, where the probability of the data given the nullhypothesis $P(D|H_0)$ is totally determined by the trend $\omega$ divided by its error $\sigma_\omega$. This is also the test statistic (cf. App. B), where it has to be noted, that the error of the trend has to be understood as the error of a mean value, hence it scales amongst others with the number of observations as can be seen in Eq. 4.8. The curve in Fig. 4.5 is

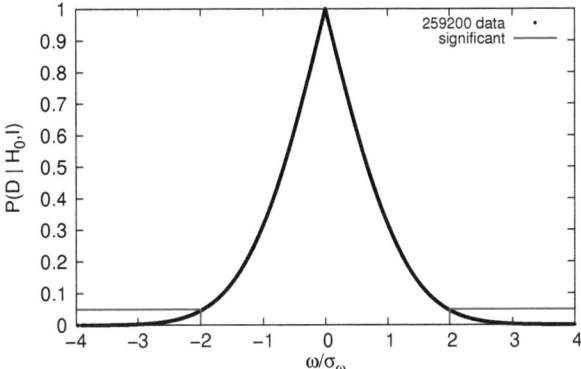

Figure 4.5: Probability of the data given the nullhypothesis $P(D|H_0)$ with $H_0 : \hat{\omega} = 0$ and $\omega$ is the water vapour trend.

not a continuous line, it is rather composed of 259200 points. Furthermore the grey lines indicate significant trends, where $P(D|H_0) < 0.05$, and this is exactly at $|\omega|/\sigma_\omega > 2$. Thus, the criterion for a statistically significant trend on a 95% confidence level is given by the claim that the absolute value of the trend has to be greater than two times its error, which is the famous and widely used standard criterion. Figure 4.6 depicts only the statistically significant trends, estimated using the frequentist statistics. In addition the significance criterion is extended by the claim, that the time series have to contain at minimum 2/3 of the maximum data points and this additional criterion is denoted with $\ell \geq 2/3T$, where $\ell$ is the number of data points of a specific time series and $T$ is the number of maximum data points. The data comprise 12 years of monthly averaged values, yielding $T = 144$ and $\ell \geq 96$. The 2/3 criterion is mainly affecting the data at the poles, where only measurements during summer are possible. Thus a few time series are not considered, where very sparse measurements are available. The significant trends are mainly strong absolute or strong relative trends. However, it is interesting that also small absolute (e.g. Antarctica) or small relative trends (e.g. Amazonia) can be significant. For instance, significantly increasing water vapour columns are found in Greenland, East Europe, Siberia and Oceania. Significant water vapour decrease is observed in the northwest USA, Central America, Amazonia, Central Africa and the Arabian Peninsula.

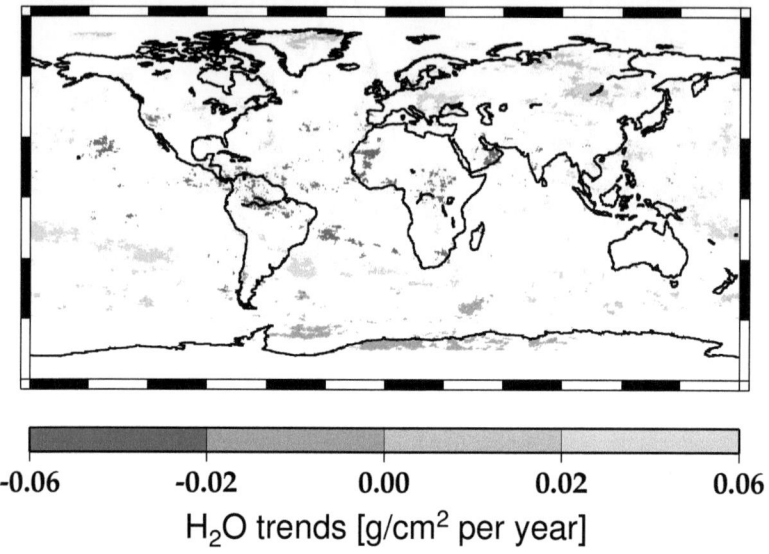

Figure 4.6: Statistically significant trends.

## 4.3 Global trend

The main result of the presented trend study is the finding of a global patchy distributed structure of positive and negative water vapour trends during the time span 1996 to 2007. These results have been made possible due to the two satellite spectrometers GOME and SCIAMACHY, which provide global water vapour measurements at a moderate spatial resolution. But regarding the observed mean global temperature increase (IPCC, 2007) an important aspect is also the global mean development of atmospheric water vapour, because of its relation to temperature (Held and Soden, 2000). Therefore the trend analysis is applied to a time series of deseasonalised globally averaged monthly mean water vapour columns for the time span from 1996 to 2007. During this time a strong ENSO (El Niño

Southern Oscillation) event took place (1997/1998). El Niño is a natural recurring (without a constant period) climate phenomenon mostly (but not solely) impacting the tropics. With respect to atmospheric water vapour the connection is performed through increasing and decreasing (depending on geolocation) surface temperatures, which cause increase and decrease of evaporation. The influence of the large El Niño event in 1997/1998 on the water vapour columns is shown in Wagner et al. (2005). Also sea surface temperature is influenced by El Niño, but trend studies by Good et al. (2007) showed, that El Niño is not influencing the trends significantly for a 20 years data record, which is a great advantage of long data sets. The GOME/SCIAMACHY data used in this study comprise only 12 years, thus the impact of El Niño on the calculated trends has to be investigated.

After the strong 1997/1998 El Niño, two small El Niño events took place in 2002 and 2006. Figure 4.7 shows the sea surface temperature (SST) anomalies (black) and the GOME/SCIAMACHY water vapour total column anomalies (grey) for the area from 4°N to 4°S and 150°W to 90°W, which are both smoothed by a 5 months running mean filter. The El Niño event in 1997/1998 exceeds the other events by a factor of about 3. This strong coupling of the near-surface temperature anomalies with the water vapour total column anomalies is also shown in Wagner et al. (2006) for GOME measurements.

As can be seen from Fig. 4.7 the two El Niño events in 2002 and 2006 are small compared to the El Niño in 1997/1998. Here, it can be benefited from the consideration of the autocorrelation during the fit routine (cf. Sect. 4.1.1), because the change in water vapour, possibly caused by an El Niño event, changes the autocorrelation of the data. For instance increasing water vapour columns over a limited time yield to systematics in the noise and therefore to increasing autocorrelation which yields to a higher error $\sigma_{\hat{\omega}}$ of the trend, because autocorrelations are considered in Eqs. (4.7) and (4.8). Hence it is not necessary to remove small events such as 2002 and 2006.

### 4.3.1 Globally averaged water vapour trend

When accumulating spatial measurements, which are gridded on a cylindrical equidistant projection (also known as Plate Carée), which is used in this study, a weighted mean has to be used, where the weights are given by the cosine of the latitude of each grid point, to account for the different surface areas. The deseasonalised globally averaged monthly mean water vapour columns are shown in Fig. 4.8 as grey filled circles connected with lines. The black line in Fig. 4.8, corresponding to the fit parameter $\hat{\omega}$, shows an increase of $0.0042 \, \text{g/cm}^2 \pm 0.0024 \, \text{g/cm}^2$ per year, i.e. 0.29% per year related to the fitted parameter $\hat{\mu} = 1.46 \, \text{g/cm}^2$. This trend is non-significant in the strict sense, where $|\omega| > 2\sigma_{\hat{\omega}}$ is required, but it is nearly

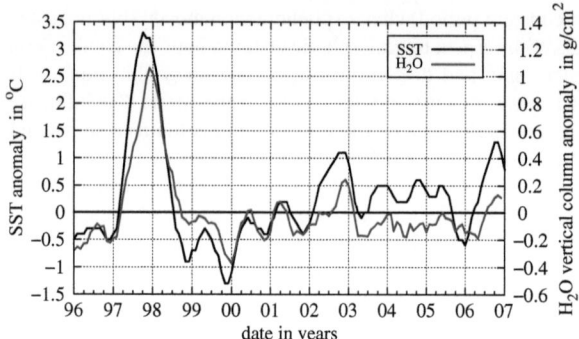

Figure 4.7: Monthly mean sea surface temperature (SST) anomalies (black) and GOME/SCIAMACHY water vapour total column anomalies (grey) averaged for the area 4°N to 4°S and 150°W to 90°W and both smoothed by a 5 months running mean filter. SST Data taken from http://coaps.fsu.edu/jma.shtml

significant. The error is strongly increased due to high autocorrelation of $\widehat{\phi} = 0.68$, which can be seen from Eq. (4.8).

One reason for the high autocorrelation is the presence of high water vapour column amounts around the year 1998, which are most likely caused by the El Niño event. These higher columns are also reported by Wagner et al. (2005) for water vapour retrieved from GOME data by a different algorithm.

### 4.3.2 Influence of El Niño 1997/1998 on the global trend

As stated above the 1997/1998 El Niño event is most likely influencing the trend in Fig. 4.8, and probably data obtained during the El Niño time have to be removed as a kind of recurring phenomenon. Otherwise it is not clear if El Niño can be totally separated from the trend, because it cannot be excluded that for instance due to an increasing water vapour trend the magnitude of the El Niño is increased. Nevertheless, the strong 1997/1998 El Niño is identified in the time series and the corresponding data are removed to quantify the effect on the trends, especially on the significance of the trends. The influence of the 1997/1998 El Niño is shown in

### 4.3 Global trend

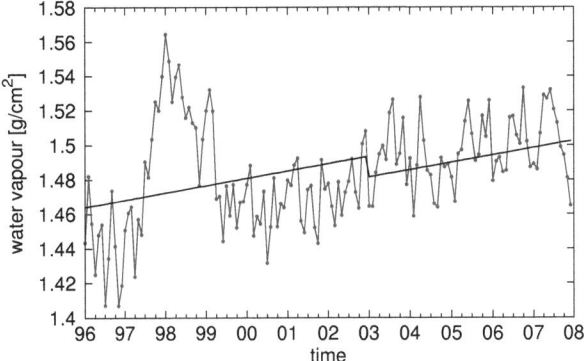

Figure 4.8: Time series of deseasonalised spatially averaged monthly means of the entire globe with the trend (black line) regarding autocorrelations.

Fig. 4.9, where the months are plotted against the years and the globally averaged deseasonalised water vapour column amounts are coded in a greyscale. As can be seen from Fig. 4.9, high water vapour columns are observed from September 1997 until March 1999. Accordingly, the global trend analysis is performed again with the data set where the potentially El Niño influenced data have been removed. Figure 4.10 shows the patterns of the significant trends. Light grey colour indicates, that here only the complete data give significant trends. Dark grey colour depicts areas, where only the data with removed El Niño time span (from September 1997 to March 1999) give significant trends. Finally medium grey colour presents locations where both data sets, with and without the El Niño time span, give significant trends. Mostly medium grey patterns are found in Fig. 4.10, which means, that in both cases (with and without El Niño) significant trends are observed, thus removing the potentially El Niño influenced data is not really needed for the data set, which is most satisfiable, because removement of data is often critical. However, for single time series, such as the globally averaged data, the El Niño influence can be crucial and removing of data points may be required.

Figure 4.11 depicts the deseasonalised spatially averaged monthly mean column amounts of the data with the El Niño event removed. The trend (black line) yields $0.0040\,g/cm^2 \pm 0.00009\,g/cm^2$ per year or $0.28\,\%$ per year thus the trend is highly

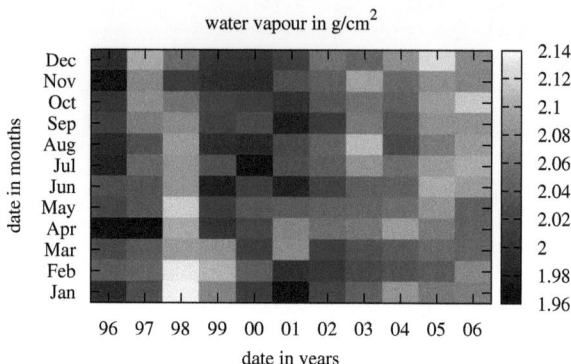

Figure 4.9: Time series of months plotted against years, while the deseasonalised globally averaged water vapour column amounts are coded with a greyscale.

significant with $\hat{\omega} > 44\sigma_{\hat{\omega}}$. Comparing Figs. 4.8 and 4.11 the trends are in principle not influenced by the El Niño data, but the errors of the trends are extremely susceptible to the El Niño data. Without the data of the El Niño time span, the autocorrelation is reduced to $\phi = 0.24$. Thus, these finding again supports the importance of implementing the autocorrelations into the regression formalism.

### 4.3.3 Water vapour correlation with temperature - Granger causality

The strong correlation between water vapour and temperature has been shown by Wagner et al. (2006) for globally averaged monthly means of GOME water vapour and temperature measurements.

This also applies to the combined globally averaged GOME/SCIAMACHY data set. Figure 4.12 shows the GOME/SCIAMACHY monthly data from 1996 to 2005 together with the globally averaged GISS (Goddard Institute of Space Studies) surface temperature data (Hansen and Lebedeff, 1992). The GISS data set is based on the Global Historical Climatology Network (GHCN), which comprises 7280 stations, the United States Historical Climatology Network (USHCN) with more than 1000 stations and the Scientific Committee on Antarctic Research (SCAR) with sta-

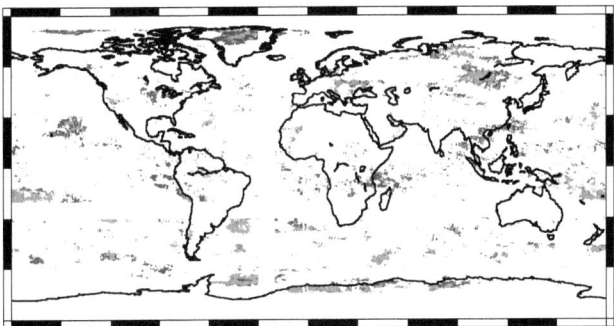

Figure 4.10: Patterns of significant trends from only the complete data (light grey), only the data with removed El Niño period from September 1997 to March 1999 (dark grey) and collocating significant trends from both (medium grey).

tions in Antarctica. A visual inspection of the two time series makes clear, that the two sets of data are correlated and both are strongly influenced by the strong El Niño in 1997/1998. Moreover, it seems to be, that the water vapour (black) is slightly ahead the temperature. To verify this impression the cross correlation function of the two variables is calculated, which is shown in Fig. 4.13. The cross correlation function shows high correlations in both directions with small positive and negative lags $\tau$. According to Granger (2001) such an observation could be an evidence for a feedback system.

Granger (1969) developed a statistical concept called Granger causality. Granger causality means, that a signal $X$ "Granger causes" (or "G-causes") a signal $Y$, if the values of $X$ can be better predicted using past information not only from $X$ itself, but also from $Y$. The use of such statistical methods in the context of prediction must be used with great care. Granger causality does not mean causal in the strict sense, especially, when not all possibilities are investigated. Therefore the results from the analysis of the Granger causality regarding water vapour and temperature should not be overemphasised. For instance Triacca (2005) showed, that the Granger causality analysis was not able to find significant results of the relationship between atmospheric carbon dioxide and temperature. However, in this approach the works from Kaufmann and Stern (1997) are followed, who found, that the

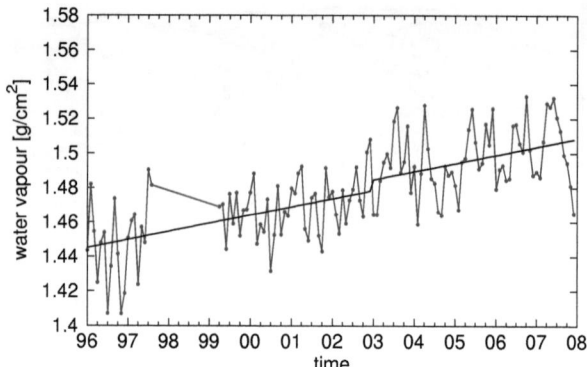

Figure 4.11: Time series (with removed El Niño measurements) of deseasonalised spatially averaged monthly means of the entire globe with the trend (black line) regarding autocorrelations.

Earth's southern hemispheric temperature G-causes the northern hemispheric temperature. The model, which describes the water vapour data is given by:

$$W_t = \mu_1 + \omega_1 X_t + \sum_{i=1}^{s} \varphi_{1i} W_{t-i} + \sum_{i=1}^{s} \gamma_{1i} T_{t-i} + \epsilon_{1t}, \quad (4.9)$$

where the $W_t$ are globally averaged deseasonalised and level shift corrected monthly mean water vapour columns. The temperature data are given by:

$$T_t = \mu_2 + \omega_2 X_t + \sum_{i=1}^{s} \varphi_{2i} W_{t-i} + \sum_{i=1}^{s} \gamma_{2i} T_{t-i} + \epsilon_{2t}, \quad (4.10)$$

where $X_t$ is the time, $\epsilon_{jt}$ are iid (independent and identically distributed) error terms and $\mu_j$, $\omega_j$, $\varphi_{ji}$ and $\gamma_{ji}$ are the regression coefficients. The length of the maximum lag $s$ is a crucial point and a good way would be to perform a Bayesian model selection. In the context of Granger causality several methods have been proposed to find the best lag length (Thornton and Batten, 1984) e.g. the Bayesian Information Criterion (BIC) also called Schwarz Information Criterion (SIC) after Schwarz (1978). The BIC is calculated in the following way:

$$BIC_j(k) = n \cdot \ln(RSS_j/n) + k \cdot \ln(n), \quad (4.11)$$

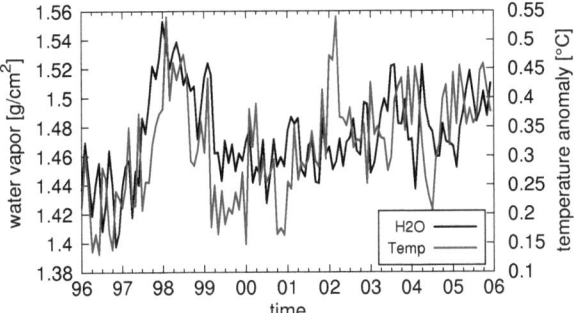

Figure 4.12: Globally averaged water vapour (black) and temperature (grey) data. The strong positive correlation can be seen clearly, especially around 1998 during the strong El Niño event.

where $n$ is the number of data points (in this case 120 minus the respective lags), $RSS$ is the sum of squared residuals between the data and the model and $k$ is the number of regression coefficients. Thus, the BIC is an increasing function of the $RSS$ and $k$, but these are coupled in opposite directions, i.e. increasing the parameters $k$ decreases the $RSS$. Accordingly the minimum of the BIC for different parameters $k$ is a compromise between small residuals and few parameters. A minimum is found at lag 1 for the sum of the BIC's from water vapour and temperature, but this function has also small values at lags 2,3 and 4. Therefore, instead of choosing only the optimal lag length (in the sense of the BIC), several lag lengths are used, which makes it possible to see the strong influence of this quantity.

Water vapour not G-causes temperature, if the $\varphi_{2i} = 0$ and temperature not G-causes water vapour, if the $\gamma_{1i} = 0$. If this is true for both, then no Granger causality exists. According to Kaufmann and Stern (1997) the significance of the Granger causality can be tested using a F-test with the test statistic:

$$w_j = \frac{(RSS_r - RSS_u)/q}{RSS_u/(n-k)}. \qquad (4.12)$$

$RSS_u$ is the sum of squared residuals of the unrestricted models Eq. 4.9 and 4.10, $RSS_r$ represents the sum of squared residuals of the restricted models, where in

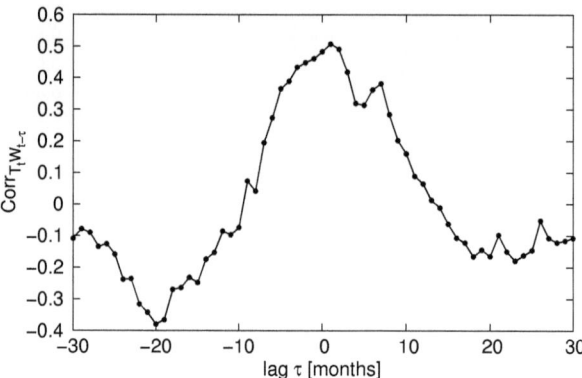

Figure 4.13: Cross correlation function of water vapour and temperature. The water vapour data are shifted with lags $\tau$ against the temperature data. High correlations are observed at small positive and negative lags $\tau$.

Eq. 4.9 the $\gamma_{1i}$ are set to zero and in Eq. 4.10 the $\varphi_{2i}$ are set to zero. $q$ is the number of coefficients restricted to zero, $n$ is the number of data points and $k$ denotes the number of regression coefficients in the unrestricted model. Strictly speaking: If the sum of the squared residuals $\epsilon_{1t}$ in Eq. 4.9 is significantly reduced through adding the information from the temperature, than the temperature G-causes the water vapour. And if the sum of squared residuals $\epsilon_{2t}$ in Eq. 4.10 is significantly reduced through adding the information from the water vapour, than the water vapour G-causes the temperature.

As said above the differences in the sums of the BIC's at lags 1,2,3 and 4 are quite small, hence these lags are used to estimate the significance for Granger causality. The results are shown in Tab. 4.1. The $p$-value, in this case, denotes the probability that the difference $RSS_r - RSS_u$ is equal zero, which means that no Granger causality exists. Regarding Tab. 4.1 the $p$-values are generally low, which supports the hypothesis of existing Granger causality. Especially for lag 1, both, water vapour G-causes temperature significantly and temperature G-causes water vapour significantly on a 95 % confidence interval. Such bidirectional Granger causality is a sign for a feedback system (Granger, 2001), which has also been insinuated by the cross correlation function (Fig. 4.13). Using higher lags than 1, the significance

| lag $s$ | $w_1$ | $p_1$ | $w_2$ | $p_2$ |
|---|---|---|---|---|
| 1 | 3.86 | 0.024 | 3.84 | 0.024 |
| 2 | 1.31 | 0.275 | 2.55 | 0.059 |
| 3 | 1.55 | 0.192 | 1.62 | 0.175 |
| 4 | 1.05 | 0.391 | 1.64 | 0.157 |

Table 4.1: Results from the analysis of Granger causality with water vapour and temperature. For different lags, the test statistic $w_j$ and their $p_j$-values are displayed. Small $p_j$-values indicate, that temperature is G-caused by water vapour ($j = 1$) and vice versa ($j = 2$).

for Granger causality is above the 5% level, however still low. In these cases the probabilities, that water vapour is G-caused by temperature are higher.

Concluding, using AMC-DOAS water vapour data and the GISS temperature data set, Granger causality is observed and temperature and water vapour are identified as a feedback system, which is in line with the understanding of atmospheric processes. This finding supports the high quality of the data. As mentioned above, the results from the Granger causality analysis should not be overemphasised. However this analysis supports the hypothesis that the feedback mechanism of temperature and water vapour is actually in full operation. It could be possible, that this mechanism may not be stopped easily in the near future. Therefore the continuation of temperature and water vapour (GOME-2) measurements is of utmost importance to monitor the ongoing climate change.

# 5 Comparison of water vapour trends

## 5.1 Intercomparison of satellite and radiosonde trends

The comparison of data or quantities such as means and standard deviations is a common problem in diverse (scientific) areas. Often, quantities are given with their respective error bars, which define certain confidence levels. If there is no overlap between the error bars it is commonly stated, that the two parameters differ significantly, which is not always valid as shown in Lanzante (2005), who recommends instead to use standard hypothesis testing such as the Gauss-test or the Student's t-test (Fahrmeir et al., 2004).

The comparison of parameters or quantities is also extensively discussed under the Bayesian statistics framework. For instance Bretthorst (1993) and Sivia and Skilling (2006) compare means and standard deviations of data using Bayesian model selection.

In the following both approaches are followed, the frequentist and Bayesian concepts, to compare the water vapour trends derived in Sect. 4 from GOME/-SCIAMACHY data with independent water vapour trends from radiosonde measurements.

As mentioned in Sect. 2.3.1 the frequentist and Bayesian concepts follow different philosophies, although similar problems can be approached. This is also the case in comparing trends. The frequentist methods can work out the probability of the data, showing same trends, assuming equal water vapour trends observed from satellite and ground measurements. This can also be interpreted as estimating the probability of an effect, assuming a known cause. The Bayesian approach gives the inverse, the probability of the cause in the case of observing certain effects. The Bayesian concept estimates the probability that the two sets of data have a combined trend, which can be interpreted as observing data from the same basic population. The especial attractiveness of the Bayesian concept lies in the fact, that it is a real model selection or hypothesis testing, which is also extensively used in decision theory (e.g. Berger (1993)). In the case of water vapour trends from two independent instruments only two hypotheses come into consideration, which are that the trends are equal or unequal. A rigorous application would prefer one hypothesis over the other, if the probability is larger than 50%. However, Jeffreys

(1939) proposed a useful scale on the evidence against hypothesis A with respect to B, which is shown in Tab. 5.1. Regarding Tab. 5.1, P(A) and P(B) denote the re-

| $\log_{10}\left(\frac{P(B)}{P(A)}\right)$ | Evidence against A |
|---|---|
| 0 to 0.5 | nothing but a chance |
| 0.5 to 1 | substantial |
| 1 to 2 | strong |
| >2 | decisive |

Table 5.1: Judgement of evidence against hypothesis A regarding Jeffreys (1939)

spective probabilities of the hypotheses and a value of $\log_{10}(P(B)/P(A)) = 1$ means, that hypothesis B is ten times more probable than hypothesis A.

## 5.2 The radiosonde water vapour data

Radiosonde water vapour measurements constitute the basis for meteorological applications and are also the calibration source for e.g. the SSM/I (Special Sensor Microwave Imager) satellite instrument (see e.g. Andersson et al. (2007)). GOME/SCIAMACHY provide an independent data set, therefore the comparison of trends with independent data from GOME/SCIAMACHY could reveal much insight and would upgrade the quality of both if agreement would be observed. The comparison will be performed with monthly averaged data from both satellites and radiosondes, which are representative for the respective months. The radiosonde data are provided by the German Weather Service (DWD) under the framework of the WMO (World Meteorological Organisation). The data set comprises 908 time series from globally distributed stations. The radiosonde data are originally available as daily vertical profiles. These have been integrated up to 100 hpa and averaged to monthly means by the DWD. As a quality criterion it is claimed that the time series have to comprise minimal 2/3 of data points to come into consideration for the comparison. As in Sect. 4.2 this constraint is used, because time series with large data gaps may not be representative. The disadvantage of the 2/3 criterion is that only 187 radiosonde time series fulfil this requirement, whereas the satellite data fulfil the 2/3 criterion in all 908 cases. This elucidates an advantage of the GOME/SCIAMACHY data, which are in principle continuously available on a monthly means basis. Nevertheless, the intercomparison is performed with the 187 collocating high quality water vapour time series.

Several differences exist between the two sets of data. The satellite spectrometers GOME and SCIAMACHY have a large footprint on Earth of about 40 km ×

320 km and 30 km × 60 km, respectively. Hence, they "see" a large area and the resulting water vapour column is an average over this area. Additionally the data are gridded on a 0.5° × 0.5° lattice. Moreover, both satellites, ERS-2 and ENVISAT, fly on a sun-synchronous orbit, which results in an equator crossing of ERS-2 at 10:30 am and that of ENVISAT at 10:00 am. Thus the GOME/SCIAMACHY time series represent strictly speaking a snap-shot climatology at a certain time for every geolocation. Further on it is a cloud adjusted climatology, because too cloudy scenes cannot be retrieved by the algorithm (Noël et al., 2004). Thus the data have to be seen in this context. On the other hand, the radiosonde stations constitute very localised point measurements. When they are compared with the satellite data, the stations lie inbetween the 0.5° × 0.5° grid points of the respective GOME/SCIAMACHY footprint. Therefore discrepancies can be well imagined. Another aspect are the measurement times of the radiosondes, which are typically at 6am, noon, 6pm and also at midnight. Moreover, some stations only provide the morning data, morning and noon data or any other permutation, thus only representing a climatology at a certain daily local time. Bias problems of radiosonde measurements have been reported by e.g. Turner et al. (2003), where peak to peak differences between radiosondes greater than 25% have been observed.

The discussion on the differences of the two data sets makes clear, that there are various reasons for discrepancies between the data. However, the comparison of trends also has an advantage against comparing e.g. the single columns. The trend should be independent from certain offsets, as long as these are constant over time. In this context, an excellent or good agreement between collocated water vapour trends represents an extremely believable verification that the trends are real. On the other hand a bad agreement does not necessarily mean, that one instrument is wrongly measuring and the other is right, or that both trends are wrong. Regarding the differences between the measurement methods, a disagreement of the collocated trends could mean, that e.g. the radiosonde captures a very localised event, which cannot be seen with the satellite.

## 5.3 Regression analysis of satellite and radiosonde data

To compare the water vapour trends from GOME/SCIAMACHY and radiosondes under the framework of frequentist statistics as well as Bayesian statistics the same trend model as in Sect. 4.1.1 is used. Thus, the GOME/SCIAMACHY data are described with

$$D^*_{1t} = \mu_1 C^*_{1t} + \omega_1 X^*_{1t} + \delta U^*_t + \epsilon_{1t}, \qquad t = 1,...,T_1, \tag{5.1}$$

where the $^*$ denotes, that the autocorrelations of the noise $N_t$ have been considered. The monthly mean water vapour columns are described by $D_{1t}^*$. The regression parameters $\hat{\mu}_1$, $\hat{\omega}_1$ and $\hat{\delta}$ have been estimated in a least square sense (cf. Sect. 4.1.1). Furthermore, the errors of the regression parameters $\sigma_{\hat{\mu}_1}$, $\sigma_{\hat{\omega}_1}$ and $\sigma_{\hat{\delta}}$ and the error of the noise $\sigma_1$ have been determined within the regression procedure. The trend model for the radiosonde measurements $D_{2t}^*$, where also autocorrelations have been considered is obtained as:

$$D_{2t}^* = \mu_2 C_{2t}^* + \omega_2 X_{2t}^* + \epsilon_{2t}, \qquad t = 1,...,T_2, \tag{5.2}$$

where no level shift is used. In the same way as above, the regression parameters $\hat{\mu}_2$, $\hat{\omega}_2$, their errors $\sigma_{\hat{\mu}_2}$, $\sigma_{\hat{\omega}_2}$ and the error of the noise $\sigma_2$ have been estimated. For the approximation of the Bayesian method, which is shown in Sect. 5.6, the error of the noise from the pooled data with a single trend is needed, which can be obtained by applying the least square regression to the pooled data $D_{pt}^* = [D_{1t}^*\,;\,D_{2t}^*]$.

$$D_{pt}^* = \mu_{p1} C_{p1t}^* + \mu_{p2} C_{p2t}^* + \omega_p X_{pt}^* + \delta_p U_{pt}^* + \epsilon_{pt}, \qquad t = 1,...,T_1+T_2, \tag{5.3}$$

where we have

$$C_{p1t}^* = \begin{cases} C_{1t}^*, & t \leq T_1 \\ 0, & t > T_1 \end{cases}, \quad C_{p2t}^* = \begin{cases} 0, & t \leq T_1 \\ C_{2t}^*, & t > T_1 \end{cases},$$

$$X_{pt}^* = \begin{cases} X_{1t}^*, & t \leq T_1 \\ X_{2t}^*, & t > T_1 \end{cases}, \quad U_{pt}^* = \begin{cases} U_t^*, & t \leq T_1 \\ 0, & t > T_1 \end{cases}.$$
(5.4)

Solving the least square regression, the error of the noise

$$\sigma_{p1} = \sqrt{\frac{1}{\ell_1-3}\sum \epsilon_{p1t}^2}, \qquad \text{with } \epsilon_{p1t} = \epsilon_{pt} \text{ if } t \leq T_1 \tag{5.5}$$

and

$$\sigma_{p2} = \sqrt{\frac{1}{\ell_2-2}\sum \epsilon_{p2t}^2}, \qquad \text{with } \epsilon_{p2t} = \epsilon_{pt} \text{ if } t > T_1 \tag{5.6}$$

under a single trend can be observed. The $\ell_i$ are the respective lengths of data $D_{1t}^*$ and $D_{2t}^*$, which are reduced by the number of fitted parameters.

## 5.4 Student's t-test applied to trends

Following the methods in the App. B, the t-test can be applied to trends. The null-hypothesis $H_0 : d = \omega_1 - \omega_2 = 0$ postulates that the difference of the two trends is equal zero, whereas the alternative hypothesis is $H_1 : d = \omega_1 - \omega_2 \neq 0$. The standard deviation of the difference $d$ is observed as

$$\sigma_d = \sqrt{\sigma_{\hat{\omega}_1}^2 + \sigma_{\hat{\omega}_2}^2}, \tag{5.7}$$

(see e.g. Welch (1947)) where $\sigma_{\hat{\omega}_i}^2$ are the respective variances of the trends $\hat{\omega}_i$. It has to be noted, that contrariwise to the derivation in the App. B, the variances of the trends in Eq. 5.7 have not to be divided by the number of data, because the $\sigma_{\hat{\omega}_i}$ are actually the standard errors of the trends, which already depend on the number of data.

The t-statistic is then given by

$$t = \frac{d}{\sigma_d} \tag{5.8}$$

Accordingly the t-distribution with

$$f = \frac{\sigma_{\hat{\omega}_1}^2 + \sigma_{\hat{\omega}_2}^2}{(\sigma_{\hat{\omega}_1}^2)^2/(\ell_1 - 3) + (\sigma_{\hat{\omega}_2}^2)^2/(\ell_2 - 2)} \tag{5.9}$$

degrees of freedom (cf. App. B) has to be integrated from $|t|$ to $\infty$. The result has to be multiplied by the factor two, because no prior information on the sign of $d$ exists, which requires a two tailed test. Finally the probability of the data given the null-hypothesis, which is $P(D_1, D_2|H_0)$ is derived. The $\ell_i$ are the respective length of the time series $D_{it}^*$ and are reduced by the number of fitted parameters, i.e. three for $D_{1t}^*$ and two for $D_{2t}^*$. The integrals of the t-distribution are typically tabulated in several high level programming languages such as Octave (http://www.gnu.org/software/octave/).

## 5.5 Bayesian model intercomparison

In the following a Bayesian method to compare trends in time series is presented, the Bayesian method can give an answer to the question if the trends determined from different/independent instruments are equal or not. The Bayesian model selection for the difference of trends is based on works of Bretthorst (1993) and Sivia and Skilling (2006) who estimate the difference of means and standard deviations between two sets of data. For this study, the method has been extended to compare trends:

To end this we set up two hypotheses:

A: Both sets have the same (unknown) trend $\omega$.

B: The two data sets have different (unknown) trends $\omega_1$ and $\omega_2$.

Note that the magnitudes of the trends do not matter. Hypothesis A is mathematically formulated with Eq. 5.3, while hypothesis B is described using Eqs. 5.1 and 5.2.

Following App. D we can start and ask for the posterior probability of the hypothesis A, given the respective data using Bayes' theorem:

$$P(A|D_1, D_2, I) = \frac{P(D_1, D_2|A, I) \cdot P(A|I)}{P(D_1, D_2|I)}, \tag{5.10}$$

where the $D_1$ and $D_2$ stand for the two data sets, A is the hypothesis and I describes certain relevant background information.

In the following the procedure from App. D is applied to the model comparison between the above defined hypotheses A and B. Since the absolute magnitudes of the parameters $p_1 = [\mu_{p1}, \mu_{p2}, \omega_p, \delta_p, \sigma_{p1}, \sigma_{p2}]$ from Eq. 5.3 are irrelevant we can use the marginalization rule Eq. D.4 and integrate them out

$$P(A|D_1, D_2, I) = \frac{\int d p_1 \, P(D_1, D_2|A, p_1, I) \cdot P(A, p_1|I)}{P(D_1, D_2|I)} \tag{5.11}$$

Assuming logical independence of the prior probabilities of the hypothesis A and the parameters $p$ we find:

$$\begin{aligned} P(A, p_1|I) &= P(A|I) \cdot P(\mu_{p1}|I) \cdot P(\mu_{p2}|I) \cdot P(\omega_p|I) \\ &\quad \cdot P(\delta_p|I) \cdot P(\sigma_{p1}|I) \cdot P(\sigma_{p2}|I) \end{aligned} \tag{5.12}$$
$$\tag{5.13}$$

In the same way as in (Eq. 5.11) the posterior for hypothesis B can be derived:

$$P(B|D_1, D_2, I) = \frac{\int d p_2 \, P(D_1, D_2|B, p_2, I) \cdot P(B, p_2|I)}{P(D_1, D_2|I)}, \tag{5.14}$$

with $p_2 = [\mu_1, \mu_2, \omega_1, \omega_2, \delta, \sigma_1, \sigma_2]$ from Eqs. 5.1 and 5.2 and

$$\begin{aligned} P(B, p_2|I) &= P(B|I) \cdot P(\mu_1|I) \cdot P(\mu_2|I) \cdot P(\omega_1|I) \\ &\quad \cdot P(\omega_2|I) \cdot P(\delta|I) \cdot P(\sigma_1|I) \cdot P(\sigma_2|I) \end{aligned} \tag{5.15}$$
$$\tag{5.16}$$

The denominator $P(D_1, D_2|I)$ is the evidence (cf. Eqs. D.4 and D.5), which is in this case obtained by:

$$P(D_1, D_2|I) = P(D_1, D_2|A, I) P(A|I) + P(D_1, D_2|B, I) P(B|I). \tag{5.17}$$

$P(A|I)$ is the prior probability that the trends of both sets of data are the same and because there is no reason to prefer either this hypothesis nor the alternative

## 5.5 BAYESIAN MODEL INTERCOMPARISON

$P(B|I)$ (that the trends are different) we assign both with the probability 0.5, thus they cancel out in the ratios given in Eqs. 5.11 and 5.14.

The prior probabilities $P(\boldsymbol{p}_1|I)$ and $P(\boldsymbol{p}_2|I)$ in Eqs. 5.11 and 5.14 can be pulled out of the integrals if they are independent from the parameters themselves, which can be realised by choosing them as bounded priors in the form of fully normalised uniform distributions:

$$P(\boldsymbol{p}_i|I) = \begin{cases} \frac{1}{p_{i\max} - p_{i\min}} & \text{If } p_{i\min} < p_i < p_{i\max} \\ 0 & \text{otherwise} \end{cases} \quad i = 1, 2, \qquad (5.18)$$

i.e. it is assumed that all $\boldsymbol{p}_i$ in the interval $[\boldsymbol{p}_{i\min}, \boldsymbol{p}_{i\max}]$ have the same probability. All prior probabilities, except the trend priors, occur in the numerator and denominator of Eqs. 5.11 and 5.14, respectively, thus they cancel out in the ratios. The priors of the pooled trend and the separate trends are chosen equally as $P(\omega|I) = P(\omega_p|I) = P(\omega_1|I) = P(\omega_2|I)$ with

$$P(\omega|I) = \begin{cases} \frac{1}{\omega_{\max} - \omega_{\min}} & \text{If } \omega_{\min} \leq \omega \leq \omega_{\max} \\ 0 & \text{otherwise} \end{cases}. \qquad (5.19)$$

These prior information constitutes the framework of the probabilistic analysis and interpretation of the results. If the range of possible trends is increased, larger differences of trends are probable and vice versa. Therefore the argumentation on the differing trends depends on the choice of the trend boundaries. Luckily, it can be benefitted from the trend study (Sect. 4.1.1), which gives information on the range of the trends. The boundaries for the three trend priors in Eq. 5.19 are chosen as $\omega_{\min} = -0.1 \, \text{g/cm}^2$ per year and $\omega_{\max} = +0.1 \, \text{g/cm}^2$ per year. These priors comprise more than 99.9% of all water vapour trends for the time span 1996 to 2007 (cf. Fig. 4.3), thus the probability space is not truncated and it is also not unrealisticly large chosen. Finally, also one trend prior cancels out in each of the ratios of Eqs. 5.11 and 5.14.

The only remaining quantities are the two likelihood functions, where the two data sets $D_1$ and $D_2$ from radiosonde and satellite measurements are completely independent:

$$P(D_1, D_2|A, \boldsymbol{p}_1, I) = P(D_1|A, \mu_{p1}, \omega_p, \delta_p, \sigma_{p1}, I) P(D_2|A, \mu_{p2}, \omega_p, \sigma_{p2}, I) \qquad (5.20)$$

and

$$P(D_1, D_2|B, \boldsymbol{p}_2, I) = P(D_1|B, \mu_1, \omega_1, \delta, \sigma_1, I) P(D_2|B, \mu_2, \omega_2, \sigma_2, I). \qquad (5.21)$$

A Gaussian likelihood is assumed such that the noise $\epsilon_{1t}$, $\epsilon_{2t}$ and $\epsilon_{pt}$ are normally distributed. The $D_1$ comprise $\ell_1$ independent measurements $\{D_{1t}\}$ and the $D_2$ comprise $\ell_2$ independent measurements $\{D_{2t}\}$, which leads to

$$P(D_1, D_2 | A, p_1, I) = \quad (5.22)$$
$$\left(\sigma_{p1}\sqrt{2\pi}\right)^{-\ell_1} \exp\left[-\frac{1}{2\sigma_{p1}^2}\sum_{t=1}^{T_1}(D_{1t} - \mu_{p1}C_{1t} - \omega_p X_{1t} - \delta_p U_t)^2\right]$$
$$\cdot \left(\sigma_{p2}\sqrt{2\pi}\right)^{-\ell_2} \exp\left[-\frac{1}{2\sigma_{p2}^2}\sum_{t=1}^{T_2}(D_{2t} - \mu_{p2}C_{2t} - \omega_p X_{2t})^2\right]$$

and

$$P(D_1, D_2 | B, p_2, I) = \quad (5.23)$$
$$\left(\sigma_1\sqrt{2\pi}\right)^{-\ell_1} \exp\left[-\frac{1}{2\sigma_1^2}\sum_{t=1}^{T_1}(D_{1t} - \mu_1 C_{1t} - \omega_1 X_{1t} - \delta U_t)^2\right]$$
$$\cdot \left(\sigma_2\sqrt{2\pi}\right)^{-\ell_2} \exp\left[-\frac{1}{2\sigma_2^2}\sum_{t=1}^{T_2}(D_{2t} - \mu_2 C_{2t} - \omega_2 X_{2t})^2\right],$$

where the *'s have been dropped due to simplicity, but still the transformed data are meant (cf. Sect. 5.3).

Due to the equality of the priors and the normalisation, the analysis simplifies to:

$$P(A|D_1, D_2, I) = \frac{\int dp_1 \, P(D_1, D_2 | A, p_1, I)}{P(D_1, D_2 | I)} \quad (5.24)$$

and

$$P(B|D_1, D_2, I) = \frac{P(\omega|I) \cdot \int dp_2 \, P(D_1, D_2 | B, p_2, I)}{P(D_1, D_2 | I)}, \quad (5.25)$$

The final posterior probabilities Eqs. 5.24 and 5.25 constitute highly complex function comprising products of more than 200 Gaussians, which have to be integrated over six and seven dimensions, respectively. Such multidimensional, complex probability density distributions have extremely small "peaks" and are exceedingly "steep", comparable to needles in a haystack, as stated colourful by Liu (2003). This means, that standard quadrature and even standard Monte Carlo integration algorithms are not sufficient to solve these integrals. Therefore a Markov Chain Monte Carlo (MCMC) method is used for integration (comprehensive information can be found e.g. in Gilks et al. (1995) and Robert and Casella (2005)), where the algorithm Differential Evolution Markov Chain (DEMC) explained by Braak (2006) has been implemented. Two factors essentially determine the precision of the method, which are the burn in phase ($bip$), i.e. the time the algorithm needs to work correctly and the number of samples ($nos$). These parameters are adjusted to achieve a precision of $\approx 0.01$ with $bip = 10^5$ and $nos = 10^5$, which constitutes an extensive computational effort. In the case of the 187 pairs of time series, this computation can be performed on a standard PC, but if e.g. two satellite data sets with more than 200000 pairs of time series have to be analysed, standard PC's are insufficient. Therefore, an analytical approximation of the Bayesian method is shown in the next section, which can be applied to large data sets on standard PC's.

## 5.6 Analytical approximation

DEMC is a sophisticated and powerful algorithm which is far beyond what is implemented in standard computational programming languages or packages. The disadvantage of DEMC is the large need of computational power. Sivia and Skilling (2006) derive an approximation for a Bayesian method, which compares means and standard deviations of data. This approximation can be adopted to the here presented method for trend comparison, which is shown in the following.

Using a quadratic Taylor series expansion of the logarithm of $A$'s likelihood function Eq. 5.20:

$$L_A = L_A(\widehat{p}_1) - \frac{1}{2} K_A' H_A K_A + \cdots \quad (5.26)$$

where $L_A = \log_e[P(D_1|A, \mu_{p1}, \omega_p, \delta_p, \sigma_{p1}, I) P(D_2|A, \mu_{p2}, \omega_p, \sigma_{p2}, I)]$ with a maximum at $\widehat{p}_1 = [\widehat{\mu}_{p1}, \widehat{\mu}_{p2}, \widehat{\omega}_p, \widehat{\delta}_p, \widehat{\sigma}_{p1}, \widehat{\sigma}_{p2}]$. The parameters $\widehat{p}_1$ are determined by the first partial derivatives $\partial L_A / \partial p_1 = 0$, which results exactly in solving a set of linear equations in a least square sense. Thus we can use the parameters $\widehat{p}_1$ estimated in Sect. 5.3.

The second term in Eq. 5.26 contains the vector $K_A' = p_1 - \widehat{p}_1$, which is shown explicitly in App. F. The entries of the $6 \times 6$ matrix $H_A$ in Eq. 5.26 are derived from the second partial derivatives of $L_A$ evaluated at $\widehat{p}_1$ which is shown in App. F.

Now we can calculate the approximated likelihood of hypothesis $A$ exponentiating $L_A$, which is

$$P_{Approx}(D_1, D_2|A, I) = \int dp_1 \, \exp(L_A) \quad (5.27)$$

$$= \int dp_1 \, \exp[L(\widehat{p}_1)] \cdot \exp\left[-\frac{1}{2} K_A' H_A K_A\right] \quad (5.28)$$

The first exponential in Eq. 5.28 is a constant and the second is a six dimensional Gaussian. Sivia and Skilling (2006) show how to integrate a $M-$ dimensional Gaussian by:

$$Z = \int \exp\left[-\frac{1}{2}x'\mathbf{H}x\right] d^M x \qquad (5.29)$$
$$= \frac{(2\pi)^{M/2}}{\sqrt{\det(\mathbf{H})}},$$

thus Eq. 5.28 becomes:

$$P_{Approx}(D_1, D_2|A, I) = \exp\left[L(\widehat{p}_1)\right] \frac{(2\pi)^{6/2}}{\sqrt{\det(\mathbf{H}_A)}}. \qquad (5.30)$$

We finally find:

$$P_{Approx}(D_1, D_2|A, I) = \qquad (5.31)$$
$$\left(\widehat{\sigma}_{p1}\sqrt{2\pi}\right)^{-\ell_1} \left(\widehat{\sigma}_{p2}\sqrt{2\pi}\right)^{-\ell_2} \exp\left[-\frac{(\ell_1 + \ell_2 - 5)}{2}\right] \frac{(2\pi)^{6/2}}{\sqrt{\det(\mathbf{H}_A)}}$$

Now we have to determine the alternative hypothesis $B$, that the time series have different trends $\omega_1$ and $\omega_2$. The procedure is in principle identical to the above derivations, using the quadratic Taylor series expansion of the logarithm of $B$'s likelihood function Eq. 5.21

$$L_B = L_B(\widehat{p}_2) - \frac{1}{2} K'_B \mathbf{H}_B K_B + \cdots. \qquad (5.32)$$

The quantities $\widehat{p}_2$, $K'_B$ and $\mathbf{H}_B$ are given in App. F. Accordingly we find

$$P_{Approx}(D_1, D_2|B, I) = \qquad (5.33)$$
$$\left(\widehat{\sigma}_1\sqrt{2\pi}\right)^{-\ell_1} \left(\widehat{\sigma}_2\sqrt{2\pi}\right)^{-\ell_2} \exp\left[-\frac{(\ell_1 + \ell_2 - 5)}{2}\right] \frac{(2\pi)^{7/2}}{\sqrt{\det(\mathbf{H}_B)}}.$$

Finally the posterior probabilities have to be normalised. Because

$$P_{Approx}(A|D_1, D_2, I) + P_{Approx}(B|D_1, D_2, I) = 1 \qquad (5.34)$$

with the evidence

$$P_{Approx}(D_1, D_2|I) = P_{Approx}(D_1, D_2|A, I) + P_{Approx}(D_1, D_2|B, I) \qquad (5.35)$$

we get

$$P_{Approx}(A|D_1,D_2,I) = \frac{P_{Approx}(D_1,D_2|A,I)}{P_{Approx}(D_1,D_2|I)} \tag{5.36}$$

and

$$P_{Approx}(B|D_1,D_2,I) = \frac{P_{Approx}(D_1,D_2|B,I) \cdot P(\omega|I)}{P_{Approx}(D_1,D_2|I)}. \tag{5.37}$$

After normalisation all terms occurring in both Eqns. 5.36 and 5.37 cancel out in the ratios, thus we have:

$$P_{Approx}(D_1,D_2|A,I) = \frac{(2\pi)^{6/2}}{\hat{\sigma}_{p1}^{\ell_1}\hat{\sigma}_{p2}^{\ell_2}\sqrt{\det(H_A)}} \tag{5.38}$$

and

$$P_{Approx}(D_1,D_2|B,I) = \frac{(2\pi)^{7/2}}{\hat{\sigma}_1^{\ell_1}\hat{\sigma}_2^{\ell_2}\sqrt{\det(H_B)}}. \tag{5.39}$$

The posterior probability of hypothesis $B$ is also proportional to the prior probability of the trends, which is chosen in the same way as in Eq. 5.19.

The analytical approximations of the posteriors can easily be applied to existing time series and are perfectly suited for analysing large data sets of 100000s of time series, which can be processed within minutes on a standard PC.

## 5.7 Application to water vapour

The presented intercomparison methods for trends in time series, i.e. the t-test and the Bayesian model selection, have been applied to existing trends from satellite and radiosonde measurements. To simplify matters in the following, it will be referred to "agreement" of trends, but actually the respective probabilities $P(D_1,D_2|H_0)$ (t-test) and $P(A|D_1,D_2)$ (Bayes) are meant. As described in Sect. 5.2 a great advantage is the independence of the data from two different sources. The trends have been calculated using the methods described in Sect. 4.1.1. For the intercomparison a quality criterion is required, i.e. both time series have to contain at minimum 2/3 monthly mean measurements over the time span from January 1996 to December 2007, i.e. at least 96 data points from maximal 144. The advantage of this constraint is, that it assures that the trends are representative for the investigated time span and less susceptible to possible outliers.

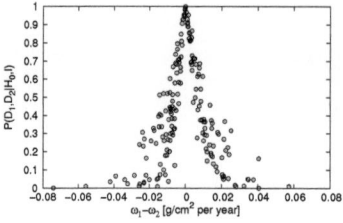

(a) t-test applied to trends. Likelihood vs. difference of trends.

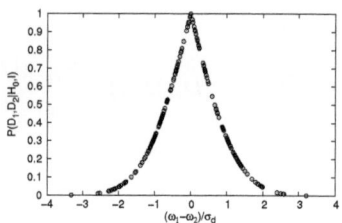

(b) t-test applied to trends. Likelihood vs. difference of trends normalised to the error of the difference.

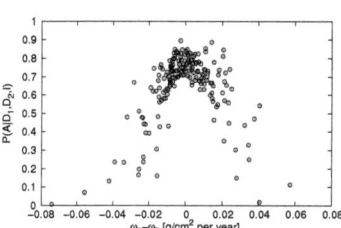

(c) Bayesian method applied to trends. Exact posterior probability vs. difference of trends.

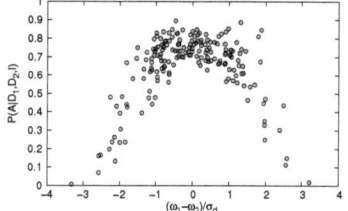

(d) Bayesian method applied to trends. Exact posterior probability vs. difference of trends normalised to the error of the difference.

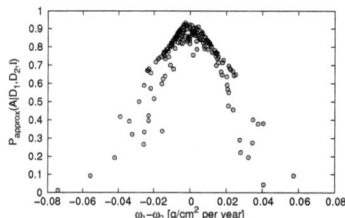

(e) Bayesian method applied to trends. Approx. posterior probability vs. difference of trends.

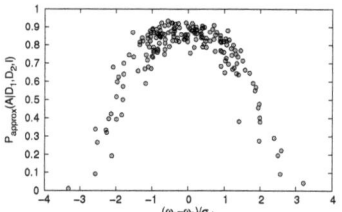

(f) Bayesian method applied to trends. Approx. posterior probability vs. difference of trends normalised to the error of the difference.

Figure 5.1: 187 probabilities of agreement between pairs of water vapour trends from GOME/SCIAMACHY and radiosonde data are plotted against the respective trend differences and against the trend differences divided by the error of the differences: (a) and (b) using the t-test, (c) and (d) applying the exact Bayesian method and (e) and (f) performing the approximated Bayesian approach.

## 5.7 Application to water vapour

Figure 5.1(a) depicts the results from the t-test: The likelihoods of the 187 trend pairs are plotted versus the difference of the trends and high probabilities are observed for small trend differences, while lower probabilities are found for large trend differences. Figure 5.1(b) shows the likelihoods plotted against the trend differences normalised to the standard deviation of the difference. From the definition of the t-test it is clear, that the probability $P(D_1, D_2|H_0)$ is totally determined by $(\omega_1 - \omega_2)/\sigma_d$, were $\omega_1$ is the radiosonde trend and $\omega_2$ is the GOME/SCIAMACHY trend. In the sense of the frequentist interpretation, statistical significance will be stated at the 95% confidence level. This means, that the null-hypothesis is rejected if $P(D_1, D_2|H_0) < 0.05$, which is true in 20 cases and it is accepted if $P(D_1, D_2|H_0) > 0.95$, which is true in 8 cases out of 187. Furthermore, the probabilities are quite equally distributed with about nine data points on average in each 0.05 probability interval. Thus, concluding it can neither be stated that the trends generally disagree nor that they do agree systematically. In the sense of the frequentist significance interpretation 159 trend comparisons, i.e. 85%, give non-significant results.

The 187 probabilities of agreement from the exact Bayesian model selection, for each trend pair, are plotted in Fig. 5.1(c) versus the difference of the trends and in Fig. 5.1(d) versus the trend difference normalised to the standard deviation of the trends (as above), $(\omega_1 - \omega_2)/\sigma_d$. Additionally the results from the approximation of the Bayesian method are shown in Figs. 5.1(e) and 5.1(f). High probabilities of agreement are found for small trend differences, whereas the probability is low for large trend deviations as in the case of the t-test. The approximation slightly overestimates the exact probabilities and the mean relative difference is in the order of 10%, but the general results from the exact method and the approximation are very similar, thus the use of the approximation can be recommended for monthly mean water vapour trend comparison, if sophisticated algorithms like DEMC are not available, large computational capacities are not accessible or large data sets have to be analysed in few time. Regarding Jeffreys' scale (Tab. 5.1 the evidence against hypothesis $B$ is substantial if the logarithm of the Bayes factor, which is here $P(D_1, D_2|A, I)/(P(D_1, D_2|B, I) \cdot P(\omega|I))$, is larger 0.5 and smaller 1, which corresponds to $0.76 < P(A|D_1, D_2, I) < 0.91$, hence the evidence against hypothesis $A$ is substantial if $0.09 < P(A|D_1, D_2, I) < 0.24$. Thus hypothesis $A$ is preferred substantially in 49 cases and $B$ in 9 cases, using the exact method. The approximation substantially prefers $A$ in 114 cases and $B$ in 5 cases. The evidence against $B$ is strong to decisive if $P(A|D_1, D_2, I) > 0.91$, which is true in zero cases for the exact solution and true in 10 cases for the approximation. Strong to decisive evidence is drawn against $A$ if $P(A|D_1, D_2, I) < 0.09$, which is observed three times in the exact case and two times in the case of the approximation. The rigorous application of the Bayesian model selection would prefer hypothesis A if $P(A|D_1, D_2, I) > 0.5$, which is true in 153 cases from 187, i.e. 82% for the exact method, and in 165 cases for the approximation.

Interpreting the observed patterns in Figs. 5.1(c) to 5.1(f), distinct clusters of data points are found between probabilities of 0.7 to 0.9. These are mostly classified as substantially supporting hypothesis A of a combined underlying trend.

In the following, examples of GOME/SCIAMACHY and radiosonde water vapour time series are analysed.

Fig. 5.2(a) shows the deseasonalised GOME/SCIAMACHY and radiosonde monthly mean water vapour columns together with their linear trends from Nottingham in England. For visual presentation the GOME/SCIAMACHY level shift has been removed. The human visual system is quite sophisticated in the identification of diverse patterns and also in comparing trends. From Fig. 5.2(a) it is clear, that the trend difference is small and indeed the trends are quite equal with $\omega_1 - \omega_2 = 0.001\,\mathrm{g/cm^2}$ per year and $(\omega_1 - \omega_2)/\sigma_d = 0.14$ (cf. Fig. 5.1). The t-test gives a probability of the data under the assumption of equal trends of $P(D_1, D_2|H_0) = 0.89$.

The Bayesian hypothesis B is visualised schematically by Fig. 5.2(a) by modelling the data with two trends. Hypothesis A is illustratively shown in Fig. 5.2(b) by pooling the data and applying a single trend. For visual presentation the offsets of GOME/SCIAMACHY and radiosonde data have been removed. From the Bayesian point of view hypothesis A is substantially preferred with $P(A|D_1, D_2, I) = 0.82$. The approximative method gives $P_{Approx}(A|D_1, D_2, I) = 0.90$. Hence, for small trend differences, both, the frequentist and Bayesian concept reveal high probabilities of agreement.

Low probabilities of agreement are found e.g. at Albany Airport in Australia. The time series are shown in Fig. 5.2(c). The visual inspection definitely classifies the trends as different. The trend difference is actually $\omega_1 - \omega_2 = 0.04\,\mathrm{g/cm^2}$ per year and after normalisation it is $(\omega_1 - \omega_2)/\sigma_d = 3.2$. The t-test gives a probability of $P(D_1, D_2|H_0) = 0.002$, the exact Bayesian finds $P(A|D_1, D_2, I) = 0.02$ and the approximation yields $P_{Approx}(A|D_1, D_2, I) = 0.04$. Thus, low probabilities are found for large trend differences by both statistical methods.

The decisive differences, as can be seen from Fig. 5.1, happen in the range between small and large trend differences. As an example, a pair of water vapour time series from Meiningen, Germany is chosen with a trend difference of $\omega_1 - \omega_2 = 0.014\,\mathrm{g/cm^2}$ per year and a normalised trend difference of $(\omega_1 - \omega_2)/\sigma_d = 1.3$.

The probability for equal trends from the t-test yields $P(D_1, D_2|H_0) = 0.19$, whereas the exact Bayesian gives $P(A|D_1, D_2, I) = 0.89$ and the approximation is $P_{Approx}(A|D_1, D_2, I) = 0.75$. Here it has again to be mentioned, that both methods (t-test/Bayes) reveal different probabilities, thus they complement each other and

## 5.7 APPLICATION TO WATER VAPOUR

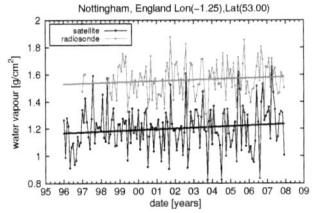

(a) GOME/SCIAMACHY deseasonalised data (black) and radiosonde data (grey).

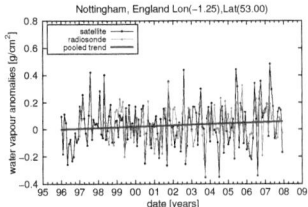

(b) Pooled GOME/SCIAMACHY and radiosonde water vapor data.

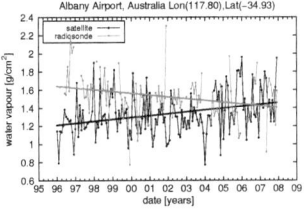

(c) GOME/SCIAMACHY deseasonalised data (black) and radiosonde data (grey).

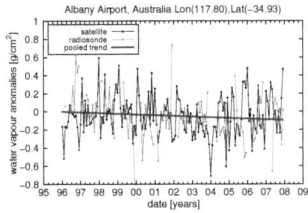

(d) Pooled GOME/SCIAMACHY and radiosonde water vapor data.

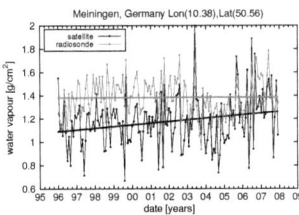

(e) GOME/SCIAMACHY deseasonalised data (black) and radiosonde data (grey).

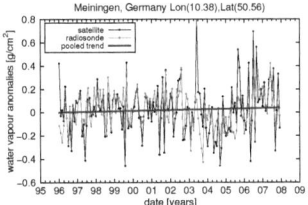

(f) Pooled GOME/SCIAMACHY and radiosonde water vapor data.

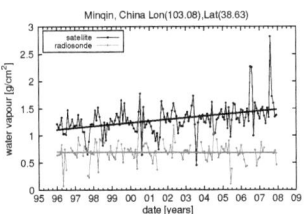

(g) GOME/SCIAMACHY deseasonalised data (black) and radiosonde data (grey).

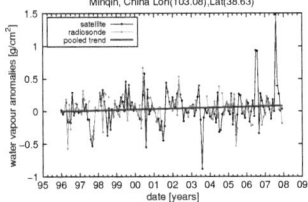

(h) Pooled GOME/SCIAMACHY and radiosonde water vapor data.

Figure 5.2: Examples of GOME/SCIAMACHY and radiosonde water vapor time series at Nottingham, England (a,b), Albany Airport, Australia (c,d), Meiningen, Germany (e,f) and Minqin, China (g,h).

are both correct under the given framework of the frequentist philosophy and the Bayesian concept. As can be seen from Fig. 5.2(e) also a visual inspection would classify the trends more different than equal, which is also reproduced by the t-test. Pooling the data, as shown in Fig. 5.2(f) makes the strong correlation between both time series clearly visible, thus the accuracy of the satellite and radiosonde measurements is strongly supported. In this sense, both methods, the t-test and the Bayesian model selection complement each other. The t-test states, that the individual time series have different trends. The Bayesian model selection gives a high probability, that both time series have a combined trend. As mentioned above this is no contradiction, because different probabilities are estimated.

The GOME/SCIAMACHY trends are plotted in Fig. 5.3, where the 187 radiosonde trends have been embedded into the figure indicated by black, grey and white bordered circles. The circles of radiosonde trends are filled with the colour for the magnitude of the respective trends according to the colour bar used also for the GOME/SCIAMACHY data. Thus, from a visual inspection, good information can be revealed, when comparing the coloured radiosonde trend-circles with the surrounding trend-colours from GOME/SCIAMACHY in the near vicinity. The borders of the circles indicate the Bayesian posterior probability $P(A|D_1, D_2, I)$ for the agreement of the trends at the specific geolocation. A black border indicates a probability of agreement of $\leq 0.5$, which means that hypothesis $B$ is preferred. It has to be noted, that 7 from 34 black bordered circles are covert by the other circles and cannot be seen in the figure. A grey bordered circle represents probabilities above 0.5 and $\leq 0.76$, where hypothesis $A$ is favoured (104 circles) and a white border indicates, that hypothesis $A$ is substantially preferred with Bayesian probabilities above 0.76 (49 circles).

One reason for discrepancies are data gaps in the radiosonde data. This has been observed e.g. at Minqin, China, shown in Fig. 5.2(g) and (f). The radiosonde data are often missing in summer, especially in 2006 and 2007, where high water vapour has been observed by SCIAMACHY. The t-test gives $P(D_1, D_2|H_0) = 0.01$, whereas the exact Bayesian gives $P(A|D_1, D_2, I) = 0.43$.

Another reasons for discrepancies between observed trends from satellite and radiosonde water vapour data is the different resolution of the two instruments. Radiosondes can capture local events, whereas the satellite measurement is an average over a large area. This will be shown in the following with an example taken from the west coast of Saudi Arabia. A blow-up of the region is depicted in Fig. 5.4, where the same colour scale is used for the GOME/SCIAMACHY and radiosonde trends as in Fig. 5.3, but with different limits. Here an increasing water vapour trend is observed with a radiosonde measurement exactly at the city of Jeddah. Also the satellite trends in the near vicinity of the town are enhanced, but not as strong as the very localised radiosonde trend. Thus, the satellite picture of

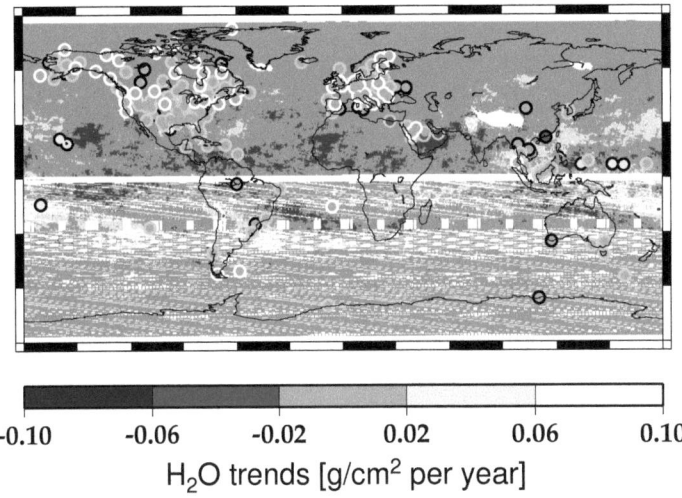

Figure 5.3: Global water vapour total column trends from GOME/SCIAMACHY are coded in greyscale colours. The 187 radiosonde water vapour trends are embedded into the figure as circles, where the same colour bar is used for the filling. White bordered circles depict Bayesian probabilities of agreements between satellite and radiosonde trends of $> 0.76$, grey borders indicate probabilities above $0.5$ and $\leq 0.76$, whereas black bordered circles show probabilities of agreement $\leq 0.5$.

the increased trends over Jeddah are more smeared out over a larger region. This explains the relative low probability of agreement between the observed trends, which is indicated by the grey bordered circle.

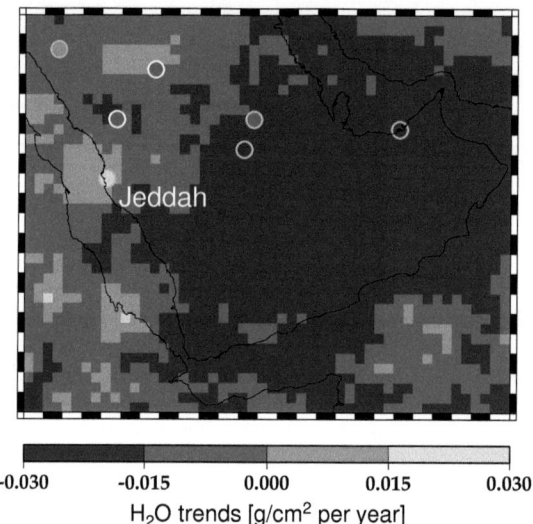

Figure 5.4: GOME/SCIAMACHY water vapour trends at the Arabian peninsular with embedded radiosonde trends. A different colour scale as in Fig. 5.3 is used in this blow-up.

# 6 Stochastic description of water vapour and temperature

## 6.1 Interaction of water vapour and temperature

The strong coupling between temperature and water vapour is well studied in laboratory experiments as performed e.g. by John Dalton in the 19th century. The dependence of water vapour on temperature can also be elucidated by the relation between the saturation vapour pressure and the temperature, based on the Clausius-Clapeyron equation. An empirical relation is given by Magnus:

$$E = 6.1\,\text{hPa} \cdot 10^{7.5T/(T+237.2\,°\text{C})} \tag{6.1}$$

Figure 6.1 depicts graphically the Magnus equation Eq. 6.1.

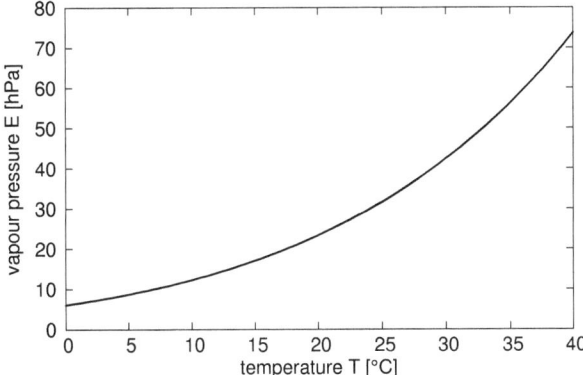

Figure 6.1: Magnus curve (cf. Eq. 6.1).

The strong correlation between water vapour and temperature has also been shown for real measurements e.g. by Wagner et al. (2006), who calculated the linear correlation coefficient for globally averaged monthly means of water vapour columns (from the GOME instrument) and temperature ($r = 0.58$), where they also used the GISS data set. In Sect. 4.3.3 the correlation coefficient between globally averaged monthly mean water vapour columns and temperature data has been estimated as $r = 0.48$, including SCIAMACHY data.

Beside these expected findings the water vapour ($H_2O$) – temperature interaction (in the following denoted HTI) in the real world is not always as simple as in laboratory experiments. Regarding the complex interactions in the Earth atmosphere from radiation over cloud condensation, wind stress and also chemical reactions it is not surprising, that also the HTI is influenced by these processes. Moreover it is clear, that temperature and water vapour interact with the surface, with e.g. vegetation, deserts, ocean and even with large cities or industrial areas. Here it is clear, that also the intervention of mankind has a potential influence on the HTI. Although these complex mechanisms are still not really understood, real measurements can incorporate these information. In this context Lenderink and Meijgaard (2008) showed, that hourly strong precipitation events occur more frequent in the Netherlands, than expected from the temperature increase and the Clausius-Clapeyron equation. These findings show, that the interplay of environmental variables cannot only be described by known physical laws, but rather has to be seen in the context of complex systems. Such results from Lenderink and Meijgaard (2008) could also be interpreted as emergent phenomena (Ebeling et al., 1998), which can be elucidated by the popular expression "the total is more than the sum of its constituents".

In 1912 A. A. Markov published his work on dependent random variables (Markov, 1912) called Markov chains. Markov chains have been used in Shannon's fundamental work on information theory (Shannon, 1948) and have undergone a renaissance in the 1970's with the increase in computational power. One of the first approaches of modelling environmental parameters with Markov chains has been performed by Waggoner and Stephens (1970), who described the succession of trees in forests. The Markovian methods have been adopted, enhanced and transferred from Isagi and Nakagoshi (1990) on plant communities, from Usher (1979) on insect populations and from Tanner et al. (1994) on coral reefs. Markov chains are also used in climate research, particularly with precipitation, e.g. describing two states "dry" and "wet" (Moon et al., 1994). A Markov chain analysis on land use in Costa Rica has been performed by Stoorvogel and Fresco (1996). Climate records from the Swiss meteorological office have been used by Nicolis et al. (1997) for a Markov chain analysis comprising three states, convective weather, advective weather and mixed weather. Nicolis and Ebeling use the Markovian analysis in the framework of complex systems (Ebeling et al., 1998) in the sense of symbolic dynamics, where they also use the terminology from statistical physics and information theory. Furthermore they show the wide range of applicability of Markov chains, e.g. the analysis of texts, sheet music and bio-sequences.

## 6.2 The Markov chain

The Markov chain is a time discrete stochastic process (Waldmann and Stocker, 2003), which obeys the Markov property: Given the present state, future states are independent from past states. If the values $i_0, ..., i_{t-1}$ of random variables $X_0, ..., X_{t-1}$ are given, then it follows, that the probability, that $X_t$ takes the value of $i_t$, depends only on $i_{t-1}$. Mathematically, Markov chains are described using conditional probabilities:

$$P(X_t = i_t | X_0 = i_0, ..., X_{t-1} = i_{t-1}) = P(X_t = i_t | X_{t-1} = i_{t-1}), \qquad (6.2)$$

which are the transition probabilities of going from $X_{t-1}$ to $X_t$. Suppose a Markov chain contains $s$ states and $X(t) \in [1, 2, ..., s]$ describes the state of a point at time $t$. If the transition probabilities are independent of $t$ (homogeneous Markov chain), then the dynamics of the Markov chain model are given by the transition matrix $\mathbf{P}$, whose elements $p_{ij}$ are the conditional probabilities:

$$P(X_t = i | X_{t-1} = j) \qquad i, j = 1, ..., s. \qquad (6.3)$$

If we denote $\pi$ as the distribution of the states of a Markov chain with the relative frequencies of the states as entries, then this distribution is recursively related to the transition matrix (Waldmann and Stocker, 2003):

$$\pi_t = \mathbf{P}\pi_{t-1} \qquad (6.4)$$

Equation 6.4 insinuates a stationary distribution, where $\pi$ is constant over time:

$$\pi = \mathbf{P}\pi, \qquad (6.5)$$

which is an eigenvalue problem with eigenvalue equal unity.

The description of environmental variables using Markov chains is valuable, but under-represented in climate time series analysis. Therefore the Markov chain analysis is applied to GOME/SCIAMACHY water vapour and GISS surface temperature, which allows the combined analysis of the two parameters. The methods are mainly based on the works of Hill et al. (2004) and Freund et al. (2006) who apply the

Markov chain analysis to a rocky subtidal community consisting of e.g. encrusting sponges and bryozoans and a marine phytoplankton community based on several diatom and flaggelate species, respectively.

The basic ideas of the application of the Markov chain analysis to environmental parameters are divided into the following steps:

1. **Preprocessing and construction of the Markov chain:** The water vapour and temperature data are reduced to discrete sequences of symbols, which represent the states of the variables at times $t$, which are encoded as high abundance, medium abundance and low abundance. Accordingly the water vapour and temperature sequences are merged together and the combinations of states (high, medium, low) form the Markov chain, which can be characterised by Eq. 6.3.

2. **Estimation of the transition probabilities:** The transition probabilities (Eq. 6.3) are estimated from the Markov chain, and the transition matrix **P** is derived.

3. **Calculation of characteristic descriptors:** Several characteristics are calculated from the transition matrix, which are e.g. *persistence, replacement of* and *entropy*, which describe the climate system. These quantities will be explained later.

## 6.3 Data sources

With the Markov chain analysis approach the interplay of water vapour and temperature will be investigated. For water vapour the global AMC-DOAS total column water vapour product is used. Several global temperature products are available, e.g. the HadCRUT3 (on a $5° \times 5°$ grid) data from the University of East Anglia or the GISS (Goddard Institute of Space Studies) surface temperatures. Here the GISS data set is used, because of its higher spatial resolution ($2° \times 2°$). The GISS data (Hansen and Lebedeff, 1992) are based on the Global Historical Climatology Network (GHCN), which comprises 7280 stations, the United States Historical Climatology Network (USHCN) with more than 1000 stations and the Scientific Committee on Antarctic Research (SCAR) with stations in Antarctica. The data sets are adjusted to the overlapping time span, which is from January 1996 to December 2005, and the GOME/SCIAMACHY data are gridded to a $2° \times 2°$ grid.

The Markov chain analysis presented here is a new approach of describing environmental parameters and does not raise the claim of a complete global analysis. Therefore the methods have been applied exemplarily to a region, where decisive environmental changes have been reported recently, i.e. the Iberian Peninsula, which is shown in Fig. 6.2. The EEA (European Environmental Agency) reports e.g. on extreme temperature changes in Europe, where the Iberian Peninsula is one of the mostly affected regions (EEA/JRC/WHO, 2008). It is also susceptible to an increase in the frequency of heat waves in summer and a decrease of frost colds in winter. Nevertheless, the Markov chain analysis can of course be applied to arbitrary regions. This will be done by analysing data from a region at the islands of Hawaii in the central Pacific Ocean. The results from the two different climate systems will be investigated under a significance analysis in Sect. 6.7.

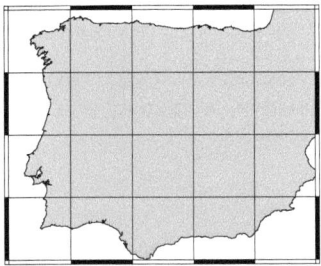

Figure 6.2: Iberian Peninsula subdivided into 20 2° × 2° grid pixels.

## 6.4 Preprocessing and construction of the Markov chains

The following analysis is performed with water vapour and temperature anomalies. The, in principle, continuous water vapour and temperature data are reduced to a sequence of discrete states. In the framework of symbolic dynamics, this approach is known as partitioning (Freund, 1996). An important aspect of the reduction of the data is the requirement of a small number of states to make the estimation of frequencies and transitions from a sample reliable. The distributions of the water vapour and temperature data from the Iberian Peninsula are shown in Fig. 6.3. The frequency distributions of the water vapour and temperature anomalies are nearly following a Gaussian normal distribution, thus it seems reasonable to divide the data into three states $X(t) \in [1,2,3]$, which are separated by the standard deviation of the data. This means, that every point $X(t) < \mu - \sigma$ is associated with state one, every point $\mu - \sigma \leq X(t) \leq \mu + \sigma$ is assigned to state two, and every point $X(t) > \mu + \sigma$ belongs to state three. The division of the data into three states, seems

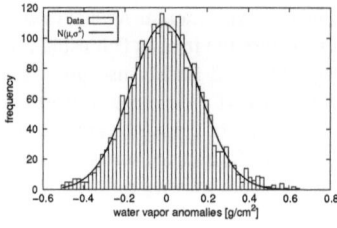 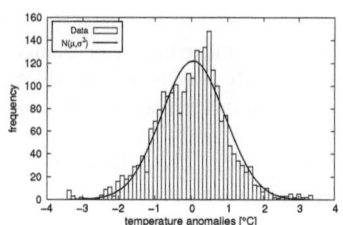

(a) Distribution of water vapour anomalies from the Iberian Peninsula.

(b) Distribution of temperature anomalies from the Iberian Peninsula.

Figure 6.3: The frequency distributions of water vapour and temperature anomalies from the Iberian Peninsula can be described with Gaussian normal distributions, which are depicted as black lines.

to be adequate and well interpretable with respect to the normal distribution. Thus state 2 corresponds to "normal" occurrences, while states 1 and 3 are rather rare events. However, other divisions can be imaginable, e.g. the reduction to only two states, divided by the mean or the splitting into three states by 1/3 quantiles. It has to be noted, that the normal distribution is at no means a necessary criterion for the Markov chain analysis. A similar division to that using the standard deviation with normally distributed data would be a separation by the respective lower 16% quantile, middle 68% quantile and upper 16% quantile. For instance, Wessel et al. (2000) partition human heart beat data into four discrete states, based on the mean value and a parameter $a$, which corresponds to a special distance from the mean value.

The construction of the three states, out of water vapour anomalies time series, is graphically shown in Fig. 6.4(a). The time series represents the data from one single grid cell in Fig. 6.2. The black horizontal lines divide the data into the three states. The resulting Markov chain is shown in Fig. 6.4(b). As can be seen, all transitions, between the three states are occurring.

In the same way the Markov chains for the temperature data are produced. In the following the water vapour and temperature chains are combined to a single Markov chain containing nine states $X(t) \in [1(11), 2(12), 3(13), 4(21), 5(22), 6(23), 7(31), 8(32), 9(33)]$, where the basis of the states (water vapour and temperature) is given in brackets. The resulting Markov chain is shown in Fig. 6.5.

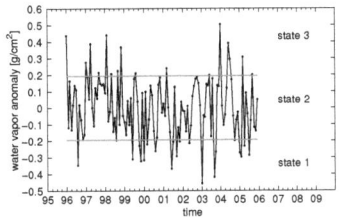
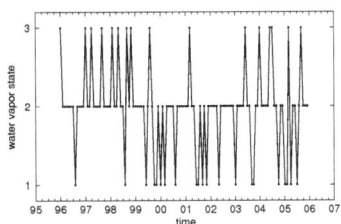

(a) Division of the time series into three "states", low values, medium values and high values.

(b) Water vapour Markov chain, consisting of three discrete states.

Figure 6.4: Reduction of a water vapour time series into a sequence of three discrete symbols (1,2,3), which can be described as a Markov chain consisting of three states.

## 6.5 Estimation of transition probabilities

As shown in the previous section, the bivariate water vapour – temperature data set has been reduced to a univariate sequence of nine discrete climate states. In this approach the assumption of homogeneity (time independence) is used and the transition probabilities are estimated over the complete time span from January 1996 to December 2005, which comprises 120 months. Furthermore spatial homogeneity is assumed over the expansion of the Iberian Peninsula, which means, that the coupling of water vapour and temperature is assumed to be the same for the entire region. The Iberian Peninsula is covered by 20 2° × 2° grid pixels (shown in Fig. 6.2), thus the analysis is based on 20 Markov chains each with 120 states for the time span from 1996 to 2005. Regarding the temporal coverage of the data (120 months), the transition probabilities have to be estimated from $120 \cdot 20 = 2400$ states or 2380 transitions between states (no transition to $t_0$).

If the absolute frequencies of the states $X(t) \in [1, 2, 3, 4, 5, 6, 7, 8, 9]$ are denoted with $f_j$, the relative frequencies are obtained by:

$$\widehat{p_j} = \frac{f_j}{\sum f_j}. \tag{6.6}$$

If $n_{ij}$ gives the number of states $j$ at time $t$, which change to state $i$ at time $t+1$, the transition probabilities are estimated with:

$$\widehat{p_{ij}} = \frac{n_{ij}}{\sum n_{ij}}. \tag{6.7}$$

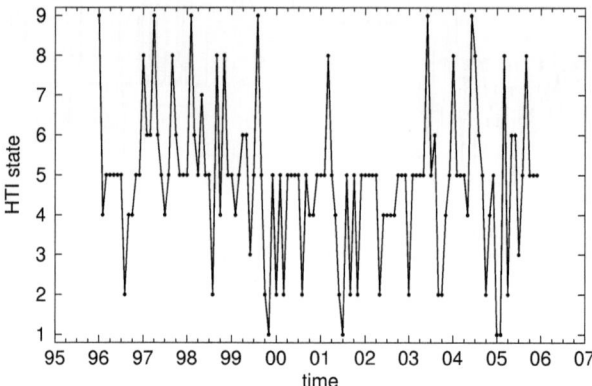

Figure 6.5: Combined Markov chain of the water vapour ($H_2O$) - temperature - interaction (HTI) with nine discrete states.

The estimated transition probabilities $\widehat{p_{ij}}$ form the transition matrix $\widehat{\mathbf{P}}$. If the matrix $\widehat{\mathbf{P}}$ is nonnegative, then it is primitive if and only if $\widehat{\mathbf{P}}^m > 0$ for some $m \geq 1$ (see e.g. Horn and Johnson (1990)). Then a Markov chain described by Eq. 6.4 will asymptotically approach an equilibrium or stationary distribution, which is given by Eq. 6.5. The transition matrix $\widehat{\mathbf{P}}$ for the Iberian Peninsula (time and space homogeneity assumed) is given in Tab. 6.1.

The transition matrix $\widehat{\mathbf{P}}$ is a stochastic matrix, which means that the sum over each column is equal unity. The entries of $\widehat{\mathbf{P}}$ are the transition probabilities $\widehat{p_{ij}}$ of changing the state from column $j$ to row $i$. Thus it is clear, that $\sum_{i=1}^{9} \widehat{p_{ij}} = 1$ for any $j$, because the probability, that the system is in one of the nine states at time $t+1$ is also unity. A striking feature of the transition matrix is obtained from the $\widehat{p_{5j}}$ (grey in Tab. 6.1), which are the highest entries for each column. This means, irrespective in which state the system is at time $t$, the probability of switching to state five is higher than changing to any other state at time $t+1$. State five corresponds to the "normal" state with medium water vapour and medium temperature. The diagonal elements $\widehat{p_{jj}}$ (bold in Tab. 6.1) denote the probability of holding state $j$ after the time step from $t$ to $t+1$. These entries are the second highest after the $\widehat{p_{5j}}$ for states two, four, five, six and eight. Thus, if the system is not going into the "normal" state five a relative high probability is observed in conserving the current state.

## 6.5 Estimation of transition probabilities

|   | 1 | 2 | 3 | 4 | 5 | 6 | 7 | 8 | 9 |
|---|---|---|---|---|---|---|---|---|---|
| 1 | **0.132** | 0.019 | 0.000 | 0.067 | 0.054 | 0.058 | 0.000 | 0.044 | 0.015 |
| 2 | 0.062 | **0.114** | 0.000 | 0.126 | 0.092 | 0.036 | 0.091 | 0.093 | 0.037 |
| 3 | 0.000 | 0.043 | **0.036** | 0.000 | 0.005 | 0.022 | 0.000 | 0.000 | 0.059 |
| 4 | 0.155 | 0.024 | 0.000 | **0.193** | 0.083 | 0.000 | 0.273 | 0.098 | 0.074 |
| 5 | 0.496 | 0.614 | 0.464 | 0.439 | **0.579** | 0.468 | 0.636 | 0.574 | 0.397 |
| 6 | 0.062 | 0.062 | 0.179 | 0.099 | 0.047 | **0.151** | 0.000 | 0.020 | 0.132 |
| 7 | 0.016 | 0.000 | 0.000 | 0.009 | 0.005 | 0.000 | **0.000** | 0.000 | 0.007 |
| 8 | 0.047 | 0.062 | 0.071 | 0.036 | 0.080 | 0.129 | 0.000 | **0.142** | 0.176 |
| 9 | 0.031 | 0.062 | 0.250 | 0.031 | 0.055 | 0.137 | 0.000 | 0.029 | **0.103** |

Table 6.1: Transition matrix $\widehat{\mathbf{P}}$, estimated from 2380 transitions in 120 months for the Iberian Peninsula. The entries $\widehat{p_{ij}}$ represent the transition probabilities of changing the state from column $j$ to row $i$.

This finding reflects the property of persistent weather conditions. Furthermore, an interesting state is the number seven, which is a strange state consisting of high water vapour and low temperature. This state most probably changes from $t$ to $t+1$ to state five or four. Thus, state seven can be interpreted as an extreme state, which is rapidly retracted to more usual states. It has to be noted, that the entire $n_{ij}$ in Eq. 6.7 are greater zero, thus the $\widehat{p}_{ij} = 0.000$ in Tab. 6.1 result from rounding.

Finally, the observed relative frequencies $\widehat{p}_j$ of the states (cf. Eq. 6.6) can be compared with the calculated equilibrium distributions of the states $\widehat{\pi}_j$, which are derived via $\widehat{\pi} = \widehat{\mathbf{P}}\widehat{\pi}$. Figure 6.6(a) shows the relative frequencies (observed (black) and calculated (grey)) of the states as a histogram, while Fig. 6.6(b) depicts a scatter plot of $\widehat{\pi}_j$ (calculated) against $\widehat{p}_j$ (observed).

The most frequent state is the number 5, the "normal" state. The fewest observed state is the "strange" state seven, but also state three, which can be interpreted as "strange" too, is quite rare.

As can be seen, the differences between the observed and estimated frequencies are marginal, thus the time and space homogeneous system of water vapour and temperature of the Iberian Peninsula has already reached the stationary distribution. In the sense of the Markov chain analysis this result is observed under the decisive assumption of temporal homogeneity, which implies a constant transition matrix over time.

# 6 STOCHASTIC DESCRIPTION OF WATER VAPOUR AND TEMPERATURE

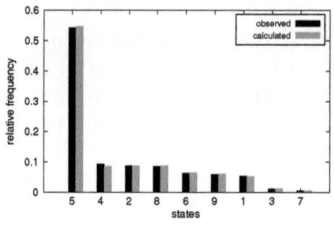

(a) Histogram of the observed ($\widehat{p}_j$) and calculated ($\widehat{\pi}_j$) relative frequencies of the states.

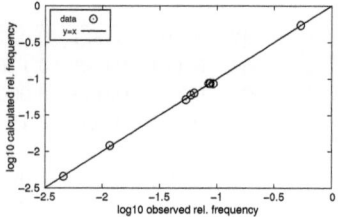

(b) Scatter plot of logarithmised (decadal) observed ($\widehat{p}_j$) and calculated ($\widehat{\pi}_j$) relative frequencies of the states.

Figure 6.6: A histogram (left) and a scatter plot (right) of the relative frequencies of the nine states indicate, that there is no difference observed between the observed (black) and calculated (grey) distributions.

## 6.6 Descriptors of the HTI

According to Hill et al. (2004) we can calculate several characteristic quantities, describing the HTI. The basis of these descriptors is given by the transition matrix of the Iberian Peninsula, for which the mean descriptors are calculated. This means, that we are not looking at a certain state and its behaviour but rather calculate averages over the states.

**Persistence.** Persistence gives the probability that the region of the Iberian Peninsula is in state $j$ at time $t$ and still in state $j$ at time $t + 1$. Thus it is a measure of the stability of a climate system.

$$P(\text{persistence}) = \sum_{j=1}^{s=9} \widehat{\pi}_j \widehat{p}_{jj} = 0.38 . \tag{6.8}$$

**Replacement of.** Replacement of is the probability that a state $j$ is changed to $i$ from $t$ to $t + 1$, i.e. the complementary event of persistence, i.e.

$$P(\text{replacement of}) = \sum_{j=1}^{s=9} \widehat{\pi}_j (1 - \widehat{p}_{jj}) = 0.62 \tag{6.9}$$

**Replacement by.** The rows of $\widehat{\mathbf{P}}$ denote the probabilities, that the state $i$ replaces other states. The average over these probabilities is $\frac{1}{s-1} \sum_{j \neq i} \widehat{p}_{ij}$, which represents the mean probability of a state to replace other states. This can be generalised for the whole system by summing over all states:

$$P(\text{replacement by}) = \sum_{i=1}^{s=9} \widehat{\pi}_i \frac{1}{s-1} \sum_{j \neq i} \widehat{p}_{ij} = 0.31 \qquad (6.10)$$

**Turnover time.** The turnover rate is the rate at which the region changes its state from $t$ to $t+1$:

$$\mathscr{T}_j = (1 - \widehat{p}_{jj}). \qquad (6.11)$$

The inverse gives the turnover time:

$$E(\text{turnover time}) = \tau_j = \frac{1}{\mathscr{T}_j}. \qquad (6.12)$$

The mean turnover time over the region is given by:

$$\overline{\tau} = \sum_{j=1}^{s=9} \frac{\widehat{\pi}_j}{\mathscr{T}_j} = 1.8 \text{ months}. \qquad (6.13)$$

**Recurrence time.** The Smoluchowski recurrence time describes the time elapsing between leaving a state $j$ and then returning to it again. The recurrence time for state $j$ is the ratio of the number of states $i \neq j$ to the number of unbroken blocks of states $i \neq j$. Kac (1947) elucidates this with an example. Suppose of having just two states $X(t) \in [0,1]$ and e.g. the following sequence of 14 observations:

$$10101010101010, \qquad (6.14)$$

then the recurrence time for state 1 is the number of 0's divided by the number of unbroken blocks of 0's, which is 7/7=1. Suppose another sequence has been observed, which is

$$100100100100100, \qquad (6.15)$$

where we find 10 zeros and five unbroken blocks of zeros, hence giving a recurrence time for state 1 as 10/5=2.

For a single HTI state the recurrence time is

$$\phi_j = \frac{1 - \widehat{\pi}_j}{\widehat{\pi}_j(1 - \widehat{p}_{jj})}. \qquad (6.16)$$

For the whole system we find

$$\overline{\phi} = \sum_{j=1}^{s=9} \widehat{\pi}_j \phi_j \tag{6.17}$$

$$= \sum_{j=1}^{s=9} \frac{1-\widehat{\pi}_j}{1-\widehat{p}_{jj}} = 9.6\,\text{months}, \tag{6.18}$$

according to Hill et al. (2004).

**Entropy:** In the context of Shannon's information theory (Shannon, 1948), the entropy is an inverse measure of predictability. The average entropy over a region is:

$$H(\mathbf{P}) = -\sum_{i=1}^{s=9} \widehat{\pi}_i \sum_{j=1}^{s=9} \widehat{p}_{ij} \log \widehat{p}_{ij}. \tag{6.19}$$

$H(\mathbf{P})$ gives the mean entropy of a region. If $H(\mathbf{P}) = 0$, then the state in the next time step is completely determined. The maximum entropy occurs under equal distribution, when $H(\mathbf{P}) = H_{\max}(\mathbf{P}) = -\log(1/s)$, i.e. the state in the next time step is completely unpredictable. Thus we can calculate a normalised entropy for the Iberian Peninsula, which is:

$$H_r(\mathbf{P}) = \frac{H(\mathbf{P})}{H_{\max}(\mathbf{P})} = 0.67 \tag{6.20}$$

The above derived descriptors can be interpreted as characteristics of a climate system. The strengths of these descriptors (which are new in atmospheric science) emerge, if different climate systems are compared among each other. Thus a different climate system of the same size as the Iberian Peninsula is analysed, which has been chosen as a region centred over the islands of Hawaii, shown in Fig. 6.7. Analogue to the derivation of the descriptors for the Iberian Peninsula, the characteristic descriptors are calculated for the Hawaiian region. The results are juxtaposed in Tab. 6.2.

To interpret these results a significance analysis is necessary, which is shown in the next section.

## 6.7 Significance of the descriptors

The basic claim on the descriptors to build the framework of a classification is that the descriptors of different climate systems are significantly unequal. Thus, we will perform a significance analysis based on simulated Markov chains on the

## 6.7 SIGNIFICANCE OF THE DESCRIPTORS

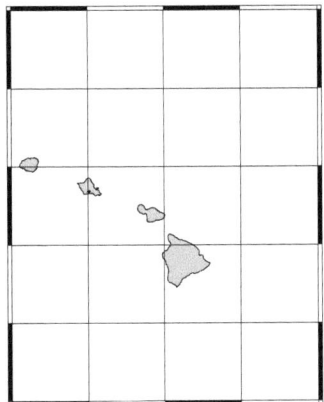

Figure 6.7: A region centred over Hawaii subdivided into 20 2° × 2° grid pixels.

|                | Iberian Peninsula | Hawaii      |
|---------------:|:-----------------:|:-----------:|
| persistence    | 0.38              | 0.44        |
| replacement of | 0.62              | 0.56        |
| replacement by | 0.31              | 0.18        |
| turnover time  | 1.8 months        | 1.9 months  |
| recurrence time| 9.6 months        | 11.7 months |
| entropy        | 0.67              | 0.63        |

Table 6.2: Markovian descriptors of two climate systems.

differences of the descriptors from the Iberian Peninsula and Hawaii. The statistical hypotheses are:

$H_0$ : There is no difference between the Iberian Peninsula and Hawaii. The observed differences between the descriptors are zero and the departure is merely due to scatter.

$H_1$ : The observed differences between the descriptors from the Iberian Peninsula and Hawaii depart from those expected by an amount, which cannot be explained by chance.

The procedure of the significance test is outlined in the following steps:

1. A pooled data set is created, consisting of the Iberian Peninsula and Hawaii data.

2. As for the separated sets of data, the pooled set is partitioned, combined to Markov chains and the transition matrix of the pooled data set is estimated.

3. Using this pooled transition matrix, 20 Markov chains of length 120 (same size as Iberian Peninsula and Hawaii data sets) are simulated twice and the descriptors are estimated for both.

4. Accordingly the differences of the simulated descriptors are calculated.

5. Steps 3 and 4 are repeated many times (here 2000 times) to achieve smooth distributions of the differences of the descriptors.

6. Finally the probability of the really observed differences can be calculated.

Exemplarily we show the normalized distribution function of the absolut differences $d_s = |p_x - p_y|$ between the persistence $p_x$ and $p_y$ estimated from two simulated data sets $x$ and $y$ of the same size as the Iberian Peninsula and Hawaii region in Fig. 6.8. Accordingly we can perform a Gauss-test and calculate the probability of observing the real difference of the persistence between the Iberian Peninsula denoted as $p_I$ and Hawaii denoted as $p_H$. The null-hypothesis is $H_0 : d = |p_I - p_H| = 0$ and the alternative is $H_1 : d = |p_I - p_H| \neq 0$ with $d = |p_I - p_H| = |0.3772 - 0.4408| = 0.0636$. The histogram in Fig. 6.8 can be described with a one sided Gaussian normal distribution $N_{\mu_d, \sigma_d}$ with $\mu_d = -1.9 \cdot 10^{-4}$ and $\sigma_d = 1.6 \cdot 10^{-2}$. The probability of observing a difference $d \geq 0.0636$ becomes:

$$P(0.0636 \leq d \leq \infty) = \int_{0.0636}^{\infty} 2 \cdot N_{\mu_d, \sigma_d}(d) \, dd \qquad (6.21)$$

$$= \int_{0.0636}^{\infty} \frac{2}{\sqrt{2\pi\sigma_d^2}} \cdot e^{-\frac{1}{2}\left(\frac{d-\mu_d}{\sigma_d}\right)^2} dd. \qquad (6.22)$$

Using the complementary error function we find:

$$P(0.0636 \leq d \leq \infty) = \mathrm{erfc}\left(\frac{d - \mu_d}{\sqrt{2\sigma_d^2}}\right) \qquad (6.23)$$

$$= \mathrm{erfc}\left(\frac{0.0636 + 1.9 \cdot 10^{-4}}{\sqrt{2 \cdot 0.016^2}}\right) \qquad (6.24)$$

$$= 9.7 \cdot 10^{-5}. \qquad (6.25)$$

## 6.7 SIGNIFICANCE OF THE DESCRIPTORS 81

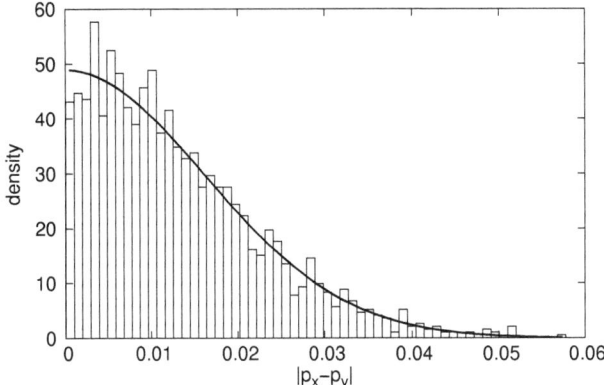

Figure 6.8: Normalized histogram of the distribution function of the differences of persistence from two simulated data sets, based on 2000 data points. The black line depicts a normal Gaussian distribution, which is fitted to the histogram.

It turns out, that the difference of the persistence between the Iberian Peninsula and Hawaii is highly significant. Making an error in rejecting the null-hypothesis of equal persistences of the two regions accounts only to a probability of $9.7 \cdot 10^{-5}$.

The significance analysis is also performed for the other descriptors. Table 6.3 summarizes the results for the intercomparison of the descriptors from the Iberian Peninsula and Hawaii. As can be seen from Tab. 6.3 all differences between the descriptors are significantly unequal zero, except the turnover time. Thus, the Markov chain analysis is able to differentiate in most cases significantly between different climate systems on the basis of measured water vapor and temperature data.

Regarding, that anomalies have been investigated, the results have to be seen relative to the seasonal cycle. Thus, if e.g. state nine, which is high water vapour and high temperature, is conserved from month $t$ to month $t+1$, then this has to be seen as a persistence of higher water vapour and temperature values than normal. The probability of persistence, which is a measure of climate stability is higher for Hawaii (0.44) than for the Iberian Peninsula (0.38), which is clear concerning the different climate zones of the two regions (tropics, mid-latitudes). The same holds vice versa for the probability of replacement of. The probability of replacement

|                | Iberian Peninsula | Hawaii         | $|d|$  | P-value          |
|---------------:|:-----------------:|:--------------:|:------:|:----------------:|
| persistence    | 0.3772            | 0.4408         | 0.0636 | $9.7 \cdot 10^{-5}$ |
| replacement of | 0.6228            | 0.5592         | 0.0636 | $9.7 \cdot 10^{-5}$ |
| replacement by | 0.3086            | 0.1850         | 0.1236 | $7.7 \cdot 10^{-12}$ |
| turnover time  | 1.8205 months     | 1.8878 months  | 0.0673 | 0.33             |
| recurrence time| 9.5713 months     | 11.6598 months | 2.0885 | $2.9 \cdot 10^{-20}$ |
| entropy        | 0.6740            | 0.6286         | 0.0454 | 0.001            |

Table 6.3: Markovian descriptors of two climate systems with the probability that the difference between the descriptors is equal zero.

by is smaller for Hawaii, which supports the more stable climatic character of the tropical island. The turnover times are a measure of the successional rates of the climate system, and the difference between the climate systems seems marginal, but is still significant in the statistical sense. The recurrence time is larger for Hawaii than for the Iberian Peninsula. On an average, the Hawaiian region, once leaves a certain state, returns to it after about 12 months, whereas the Iberian Peninsula region turns back to a state, which was left, after about 10 months. The entropy, which is an inverse measure of the predictability of a state at time $t+1$ if the state at time $t$ is known is 0.67 for the Iberian Peninsula climate system. The entropy of the Hawaiian region is significantly smaller with 0.63. However the predictability of states from one month to the next, for both climate systems, is very uncertain.

Concluding, significant different magnitudes of the descriptors for the two climate systems, the Iberian Peninsula and Hawaii, have been found. As an outlook, the incorporation of more climate parameters such as clouds, precipitation and vegetation into the Markov chain could give a more complete picture of the climate states of a region. On the basis of the Markovian descriptors it could also be possible to develop a new climate classification scheme.

# 7 Conclusions and outlook

In this study, three main points concerning atmospheric water vapour, retrieved from satellite data, have been addressed. First, a comprehensive trend analysis has been performed with global GOME/SCIAMACHY water vapour data, comprising the time span from 1996 to 2007 on a monthly mean basis. One requirement for the trend analysis has been the connection of the GOME and SCIAMACHY data sets, which has been achieved by introducing a level shift to merge the two sets of data together during the least square regression procedure. Furthermore the influence of El Niño and near surface temperature on the water vapour has been investigated.

The second part addressed the validation of the GOME/SCIAMACHY trends with independent water vapour trends retrieved from globally distributed radiosonde stations. It has been shown, that the intercomparison of trends using frequentist methods and Bayesian model selection can complement each other. However more information could be revealed from the Bayesian approach, which mostly supports the hypothesis of an underlying same water vapour trend from both instruments.

Finally, a new concept to describe the water vapour - temperature - interaction has been applied in the form of a stochastic Markov chain analysis. New descriptors have been derived, which are significantly characteristic for the climate systems under consideration. These descriptors can give information of e.g. the stability of a climate system, the successional times of changes or the short term predictability.

In the following, conclusions are drawn in more detail about the three major parts of this study.

The trend analysis (of global monthly mean water vapour data from 1996 to 2007) focussed on the estimation of the statistical significance of the observed trends. First the trends have been calculated from monthly mean water vapour column amounts where the seasonal component has been removed. Special emphasis has been placed on the consideration of autocorrelations in the data. The trend calculation, which is based on the well known least square linear regression, provides an error for the trend. This error is influenced by the length of the time series, the noise, the autocorrelation of the noise, and the level shift between GOME and SCIAMACHY data.

The significance of the trends has been estimated using the frequentist approach in form of a Gauss-test, which has revealed significant trends on a 95% level.

For the period from January 1996 to December 2007 significant increase in the water vapour columns up to 5 % per year has been found in Greenland, East Europe, Siberia and Oceania, and significant decrease up to 5 % per year has been observed in the northwest USA, Central America, Amazonia, Central Africa, and the Arabian Peninsula. The significant trends can be interpreted as tracers of the climate state, hence these regions could change their states, e.g. from dry to humid or from moist to dry. However long-term oscillations cannot be excluded.

For the whole globe an increasing trend of $0.0042\,g/cm^2 \pm 0.0024\,g/cm^2$ per year, i.e. 0.29 % per year has been observed. This trend is non-significant in the strict sense when taking into account the 1997/1998 El Niño event, which is seen in the globally averaged data as strong enhanced water vapour columns from September 1997 to March 1999. Masking out the El Niño time span – which should be done in this case – a significant water vapour trend of similar size, namely $0.0040\,g/cm^2 \pm 0.00009\,g/cm^2$ per year or 0.28 % per year has been observed. The complete global time series is strongly autocorrelated, with a magnitude of 0.6 at lag one. These high autocorrelations can be mainly attributed to the El Niño event in 1997/1998, because the autocorrelations in the data without El Niño times are reduced to 0.2. Strong autocorrelations in time series can be misinterpreted as trends by the use of simple least square regression, thus they have been considered during the regression procedure. Taking autocorrelations into account, the same trends have been observed for both time series, but different errors of the trends have been estimated, because the consideration of autocorrelations increases the errors. Hence, the consideration of autocorrelations is necessary to reveal convincing trends and errors by using least square trend estimation, especially for short time series and strong autocorrelations.

As also shown by Wagner et al. (2006), the globally averaged water vapour columns correlate with the near surface temperature. Beside the simple correlation the concept of Granger causality, which can give insights into a causal relation between the two quantities water vapour and temperature has been applied. It has turned out, that both, water vapour G-causes temperature and temperature G-causes water vapour, which is a characteristic of a feedback system. And indeed, water vapour and temperature constitute a feedback system (Held and Soden, 2000). Thus, the detection of a feedback in the relatively short water vapour and temperature time series supports the high quality of the data.

The water vapour trends derived from satellite data have been compared with water vapour trends from globally distributed radiosonde measurements. The intercomparison of the trends from independent instruments has been performed in a statistical sense using hypothesis testing. The standard frequentist approach, i.e. the well known Student's t-test has not given definite answers. Most of the pairs of trends (85 %) are evaluated as non-significant, thus neither agreement nor disagreement could be stated.

The Bayesian approach has inferred the posterior probabilities for the two hypotheses:

- $A$: Both data sets have the same trend,
- $B$: The two time series have different trends,

which is not possible using frequentist statistics hypothesis testing.

The great attractiveness of the Bayesian method is, that it is a real model selection procedure. Thus it is possible to decide between one of the two hypotheses $A$ and $B$. In the case of water vapour trends from satellite and radiosonde data, hypothesis $A$ is favoured in 82 % of the trend pairs. The evidence for hypothesis $A$ is even substantial in 26 % of pairs of trends.

However, the interpretation of agreement and disagreement has also to be seen in the context of the origin of the data as discussed in Sect. 5.2, where several possibilities for disagreement can be imagined. The two most important aspects for disagreeing trends from satellite and radiosonde measurements are constituted by missing data in the radiosonde time series and the different resolution of the two instruments. The problem of data gaps in the radiosonde data has been shown for a time series at Minqin, China. Here, water vapour measurements are mostly missing in summer, which introduces a strong bias in the trend measured by radiosonde. Thus, the advantage of in principle continuous monthly mean data from satellite is pointed out. The problem of different spatial resolutions was shown e.g. for a time series at the city of Jeddah, where the radiosonde, which is located directly in the city, can capture local events, whereas the satellite has not such a high resolution. However, the city of Jeddah is also seen as a spot of increasing water vapour in the satellite data, but not as strong as seen by the radiosonde.

The total water vapour column is strongly connected with near surface temperature. Therefore an analysis of the combined water vapour and temperature data has been applied in the form of a stochastic Markov chain analysis.

The Markov chain approach reduces the bivariate time series to a univariate sequence of symbols, representing the water vapour and temperature interaction in the form of nine discrete states, which can be described as Markov chains. As an example the Iberian Peninsula is investigated, because this region is strongly susceptible to climate change, which has been reported by e.g. EEA/JRC/WHO (2008). Assuming time and space homogeneity of the Markov chains from the Iberian Peninsula several mean characteristic descriptors have been estimated, which are new in atmospheric research. These characteristics, which are statistically significant, describe e.g. the stability of a climate system, the successional changes of water vapour – temperature interaction on time scales of months or e.g. the entropy of a system, which is an inverse measure of short term predictability. The strength of the derivation of these descriptors is the possibility of comparing different climate systems. This has been done exemplarily for the Iberian Peninsula and the Hawaii islands, which are of course different climate systems, because the first is located at mid-latitudes and the last is located in the tropics. The Markov chain analysis has revealed significant differences between the two regions in the form of the characteristic descriptors. It has turned out, that the water vapour – temperature interaction at the Hawaiian region is more stable than at the Iberian Peninsula. This has been revealed by calculating the persistence, which is the probability of conserving a certain water vapour – temperature state from one month to the next. This probability is 0.38 for the Iberian Peninsula and 0.44 for the Hawaii islands, and this difference is statistically highly significant. Another descriptor, which is characteristic for the successional changes in the system is, e.g. the replacement by. This quantity gives the probability, that a state replaces another and has been estimated to 0.31 for the Iberian Peninsula and 0.18 for Hawaii. Furthermore, the recurrence time, which represents the time elapsing between leaving a state and returning to it again and thus also constitutes a measure of succession of states in the climate system gives 9.6 months for the Iberian Peninsula and 11.7 months for the Hawaii islands. These findings are in line with the known general meteorological properties of the tropical Hawaii region (more stable) and the mid-latitudinal region of the Iberian Peninsula (more unstable). As an outlook, the Markov chain analysis offers a new possibility for a climate classification. The incorporation of other important climate variables, such as clouds, precipitation and vegetation would give a more complete picture of the climate state of a region. Using the Markovian descriptors the development of a new climate classification scheme would be a valuable continuation of this work.

The water vapour column is changing, which is derived from satellite data and validated with radiosonde measurements. The human impact on this is not clear, though the anthropogenic intervention in nature is beyond all question. On the one hand humans irrigate fields (which has a direct effect on the atmospheric water vapour columns reported by Boucher et al. (2004)) for agriculture, on the other hand they drain swamps. Woods are deforested and grassland is concreted. Diamond (2005) refers to drastic anthropogenic interventions such as deforestation and high consumption of groundwater in the northwest USA (especially in Montana), where significant water vapour decrease is detected. For instance Gordon et al. (2005) attribute a decrease in water vapour flow of the Brazilian Amazon region to 15 % deforested rainforest, which is in line with the observed decreasing trends.

Thus, a continuation of water vapour measurements is absolute necessary to monitor the ongoing climate change, which has been successfully initiated with the start of the first GOME-2 instrument on MetOp (Noël et al., 2008). One can imagine that at a certain length of the time series a simple linear regression is not suitable and eventually more complex models are needed. For instance Dose and Menzel (2006) perform a Bayesian model selection with a constant model, a simple linear trend model and a model that allows two trends on tree blossom time series, which could be a useful method for the analysis of the extended data set comprising GOME, SCIAMACHY and GOME-2 measurements.

Moreover, an extension of the Markov chain analysis with other important climate parameters, e.g. clouds, precipitation and vegetation constitutes a valuable method to detect climate change in a more complex context of the interaction of climate variables.

# A Derivation of the error of a trend

Following Fahrmeir et al. (2004) we want to minimise the function

$$\sum_{i=1}^{n}(Y_i - \alpha - \beta x_i)^2 \to \min, \tag{A.1}$$

where $Y_i$ contains the data, $\alpha$ and $\beta$ are the regression coefficients and $x_i$ represents the time. The least square estimator of the trend is given by

$$\hat{\beta} = \frac{\sum_{i=1}^{n}(x_i - \bar{x})(Y_i - \bar{Y})}{\sum_{i=1}^{n}(x_i - \bar{x})^2}. \tag{A.2}$$

The hat over $\beta$ means, that this quantity is estimated and thus error-prone. The bar over e.g. $x$ denotes the arithmetic mean.

After expanding the numerator we obtain

$$\hat{\beta} = \frac{\sum_{i=1}^{n}(x_i - \bar{x})Y_i}{\sum_{i=1}^{n}(x_i - \bar{x})^2} - \frac{\sum_{i=1}^{n}(x_i - \bar{x})\bar{Y}}{\sum_{i=1}^{n}(x_i - \bar{x})^2}, \tag{A.3}$$

and with $\sum_{i=1}^{n}(x_i - \bar{x})\bar{Y} = 0$ we find that

$$\hat{\beta} = \frac{\sum_{i=1}^{n}(x_i - \bar{x})Y_i}{\sum_{i=1}^{n}(x_i - \bar{x})^2} = \sum_{i=1}^{n}\left(\frac{(x_i - \bar{x})}{\sum_{i=1}^{n}(x_i - \bar{x})^2}\right)Y_i = \sum_{i=1}^{n}b_i Y_i, \tag{A.4}$$

with

$$b_i = \frac{(x_i - \bar{x})}{\sum_{i=1}^{n}(x_i - \bar{x})^2}. \tag{A.5}$$

From Eq. A.4 we see, that $\hat{\beta}$ is a linear function of $Y_i$.

The least square estimators are denoted as $\hat{Y}_i = \hat{\alpha} + \hat{\beta}x_i$ and the residuals are $\hat{\epsilon}_i = Y_i - \hat{Y}_i$. The mean sum of squared residuals, which is the variance, is then obtained by $\hat{\sigma}^2 = \frac{1}{n-2}\sum_{i=1}^{n}\hat{\epsilon}_i^2$, ($n-2$, because of the two fit parameters $\alpha$ and $\beta$).

To estimate the error of the trend, we assume normally distributed residuals: $\epsilon_i \approx N(0, \sigma^2)$ and also normally distributed data $Y_i \approx N(\alpha + \beta x_i, \sigma^2)$, and because $\widehat{\beta}$ is a linear function of $Y_i$, $\widehat{\beta}$ is also normally distributed:

$$\widehat{\beta} \approx N(\beta, \sigma_{\widehat{\beta}}^2). \tag{A.6}$$

$\sigma_{\widehat{\beta}}^2$ represents the variance of $\widehat{\beta}$, which can be derived using Eq. A.4:

$$Var(\widehat{\beta}) = Var\left(\sum_{i=1}^{n} b_i Y_i\right) = \sum_{i=1}^{n} b_i^2 \cdot Var(Y_i) \tag{A.7}$$

with

$$b_i^2 = \frac{(x_i - \overline{x})^2}{\sum_{i=1}^{n}(x_i - \overline{x})^2 \cdot \sum_{i=1}^{n}(x_i - \overline{x})^2} \tag{A.8}$$

$$\tag{A.9}$$

we find

$$Var(\widehat{\beta}) = \frac{\sum_{i=1}^{n}(x_i - \overline{x})^2 \cdot Var(Y_i)}{\sum_{i=1}^{n}(x_i - \overline{x})^2 \cdot \sum_{i=1}^{n}(x_i - \overline{x})^2}. \tag{A.10}$$

The $Var(Y_i)$ can be pulled out of the sum in Eq. A.10, because it is the same for all $i$. Equation A.10 can be simplified, and we find:

$$Var(\widehat{\beta}) = \frac{Var(Y_i)}{\sum_{i=1}^{n}(x_i - \overline{x})^2}. \tag{A.11}$$

With

$$Var(Y_i) = E(Y_i - (\alpha + \beta x_i))^2, \tag{A.12}$$

where $E$ denotes the expectation value and the least square estimates $\widehat{Y}_i = \widehat{\alpha} + \widehat{\beta} x_i$ it follows, that

$$\widehat{Var}(Y_i) = E(Y_i - (\widehat{\alpha} + \widehat{\beta} x_i))^2 = E(Y_i - \widehat{Y}_i)^2 = E(\widehat{\epsilon}_i)^2 = \frac{1}{n-2}\sum_{i=1}^{n}\widehat{\epsilon}_i^2 = \widehat{\sigma}^2. \tag{A.13}$$

Using Eqs. A.13 and A.11 we find the estimator for the variance of $\widehat{\beta}$:

$$\sigma_{\widehat{\beta}}^2 = \frac{\widehat{\sigma}^2}{\sum_{i=1}^{n}(x_i - \overline{x})^2}. \tag{A.14}$$

The error of $\widehat{\alpha}$ can be derived in an analogues way. In the App. C the error estimation of the trend is shown for the multivariate case in matrix notation, where also autocorrelations and a level shift are considered.

# B Student's t-test

The other important aspect of statistics used in this thesis is the hypothesis testing, especially the Student's t-test.

The t-Test is based on the Student distribution, which was developed 1908 by W. S. Gosset under his pseudonym "Student". Gosset discovered, that the standardised mean of normally distributed data follow his Student's distribution if the variance is not known and has to be estimated from the data. The corresponding probability density distribution is given as

$$\Phi_n(x) = \frac{\Gamma\left(\frac{n+1}{2}\right)}{\sqrt{n\pi}\,\Gamma\left(\frac{n}{2}\right)} \left(1 + \frac{x^2}{2}\right)^{\frac{-n+1}{2}}, \tag{B.1}$$

where n depicts the degrees of freedom of the data input x and $\Gamma$ is the Gamma-function, which is typically tabled in standard statistical and mathematical software packages.

The standard t-test is typically used for the comparison of two sets of data $D_{1t}$ and $D_{2t}$, which can be described by their mean values $\mu_1, \mu_2$ and their noise in form of the standard deviations $\sigma_1$ and $\sigma_2$, which are iid (independent and identically distributed) Gaussian noise. Accordingly one is interested in the two hypotheses whether the means are the same or they are different (within their uncertainties).

The procedure is straight forward: First the difference of the two means is calculated, which is

$$d = \mu_1 - \mu_2, \tag{B.2}$$

then the nullhypothesis can be set up, that the data are the same, if the difference is equal zero: $d = 0$. The nullhypothesis is mathematically formulated as $H_0 : d = 0$ and the alternative is $H_1 : d \neq 0$. The philosophy behind the frequentistic hypothesis testing is, that a certain cause is assumed (nullhypothesis) and accordingly the data are tested, if they are agreeable with the cause.

Following, the standard deviation of the difference $d$ has to be estimated, which is:

$$\sigma_d = \sqrt{\frac{\sigma_1^2}{N_1} + \frac{\sigma_2^2}{N_2}}, \tag{B.3}$$

where the $N_i$ are the respective length of the data (i.e. the number of data). Subsequently the t-statistic can be set up, i.e.:

$$t = \frac{d}{\sigma_d}. \tag{B.4}$$

Now one has to perform the significance test to either confirm or reject the nullhypothesis on a definite confidence level. The above t-statistic follows a t-distribution with $f$ degrees of freedom, which are calculated as follows:

$$f = \frac{\sigma_1^2/N_1 + \sigma_2^2/N_2}{(\sigma_1^2/N_1)^2/(N_1-1) + (\sigma_2^2/N_2)^2/(N_2-1)}, \tag{B.5}$$

which is called the Welch-Satterthwaite equation. Now, one has to look at the t-distribution with $f$ degree of freedom and reveal the likelihood of observing the value $t$ from Eq. B.4. For $f \to \infty$ the t-distribution converges to the standard normal distribution. Usually, for $f > 30$ a normal distribution is a very good approximation (Fahrmeir et al., 2004).

# C Trend estimation in matrix notation

The following steps show the calculation of the trend $\omega$ and the uncertainty of the trend $\sigma_\omega$ regarding autocorrelations. Equation (4.1) can be cast into compact matrix notation

$$\mathbf{Y} = \mathbf{X}\beta + \mathbf{N}, \tag{C.1}$$

where $\mathbf{Y}$ is the $\ell \times 1$ vector of observation, $\mathbf{X}$ is a $\ell \times 3$ matrix consisting of the constant $C_t$, time $X_t$ and step function $U_t$. $\beta = (\mu, \omega, \delta)'$ represents the vector of unknown regression coefficients and $\mathbf{N}$ is the noise vector afflicted with autocorrelations.

The $N_t$ are directly calculated from the time series (cf. Eq. (4.3)) and with the connection to the $\epsilon_t$ from Eq. (4.2) only the $\epsilon_t$ for $t = 1, ..., T$ can be calculated via

$$\epsilon_t = N_t - \phi N_{t-1} \tag{C.2}$$

because no $N_{-1}$ exists. Therefore the $\epsilon_0$ has to be estimated by $\epsilon_0 = \sqrt{1-\phi^2} N_0$ which is motivated by the assumption

$$\frac{\sigma_\epsilon}{\sigma_N} \approx \frac{\epsilon_t}{N_t}. \tag{C.3}$$

A matrix $\mathbf{P}'$ is constructed which satisfies:

$$\mathbf{P}'\mathbf{N} = \epsilon \tag{C.4}$$

which is in detail:

$$\begin{pmatrix} \sqrt{1-\phi^2} & 0 & 0 & \cdots \\ -\phi & 1 & 0 & \cdots \\ 0 & -\phi & 1 & \cdots \\ \vdots & \vdots & \vdots & \vdots \end{pmatrix} \cdot \begin{pmatrix} N_0 \\ N_1 \\ N_2 \\ \vdots \end{pmatrix} = \begin{pmatrix} \epsilon_0 \\ \epsilon_1 \\ \epsilon_2 \\ \vdots \end{pmatrix} \tag{C.5}$$

so that $\mathbf{N} = \mathbf{P}'^{-1}\epsilon$.
The model Eq. (C.1) becomes:

$$\mathbf{Y} = \mathbf{X}\beta + \mathbf{P}'^{-1}\epsilon. \tag{C.6}$$

Using matrix algebra, the model can be written as

$$\mathbf{P'Y} = \mathbf{P'X}\beta + \epsilon \qquad (C.7)$$

or using the transformed variables $\mathbf{Y^*} = \mathbf{P'Y}$ and $\mathbf{X^*} = \mathbf{P'X}$ we have

$$\mathbf{Y^*} = \mathbf{X^*}\beta + \epsilon. \qquad (C.8)$$

Now we have absorbed the autocorrelations in the transformed variables $\mathbf{Y^*}$ and $\mathbf{X^*}$ of model Eq. (C.8) and we can apply a least square fit. The least square estimator can be calculated by:

$$\hat{\beta} = (\mathbf{X^{*'}X^*})^{-1}\mathbf{X^{*'}Y^*}. \qquad (C.9)$$

Denoting the diagonal elements of $(\mathbf{X^{*'}X^*})^{-1}$ with $v_j$ the variance of $\hat{\beta}$ becomes:

$$\mathrm{Var}(\hat{\beta}_j) = \sigma_\epsilon^2 v_j, \qquad j = 1, 2, 3, \qquad (C.10)$$

where $\sigma_\epsilon^2$ stands for the variance of the $\epsilon_t$. Therefore the variance of the trend estimator $\hat{\omega}$ is

$$\sigma_{\hat{\omega}}^2 = \mathrm{Var}(\hat{\omega}) = \sigma_\epsilon^2 v_2. \qquad (C.11)$$

The variance $\sigma_{\hat{\omega}}^2$ or the standard deviation $\sigma_{\hat{\omega}}$, respectively, of the trend estimator considers the length of the data ($\ell$), the contained noise ($\sigma_\epsilon$), the autocorrelation of the noise ($\phi$) and additionally the position of the level shift ($\vartheta$), but not its magnitude. Thus $\sigma_{\hat{\omega}}$ can be written as

$$\sigma_{\hat{\omega}} = \frac{\sqrt{12}\,\sigma_\epsilon}{(1-\phi)\cdot[\ell(\ell^2-1)]^{\frac{1}{2}}} \cdot \frac{1}{[1-3\vartheta(1-\vartheta)]^{\frac{1}{2}}}, \qquad (C.12)$$

where $\vartheta = T_0/T$ is the fraction of the data before the level shift occurs. $T$ denotes the maximal number of data and $T_0$ represents the position of the level shift. With the assumption $\ell(\ell^2-1) \approx \ell^3$ Eq. (C.12) can be written as:

$$\sigma_{\hat{\omega}} \approx \frac{\sqrt{12}\,\sigma_\epsilon}{(1-\phi)\cdot\ell^{\frac{3}{2}}} \cdot \frac{1}{[1-3\vartheta(1-\vartheta)]^{\frac{1}{2}}}. \qquad (C.13)$$

The variance $\sigma_N$ of the autocorrelated noise $N_t$ is directly related to the variance $\sigma_\epsilon$ of the white noise $\epsilon_t$ by

$$\sigma_N^2 = \frac{\sigma_\epsilon^2}{(1-\phi^2)}, \qquad (C.14)$$

thus an approximation is found with

$$\sigma_{\hat{\omega}} \approx \frac{\sqrt{12}\,\sigma_N}{\ell^{\frac{3}{2}}} \cdot \sqrt{\frac{1+\phi}{1-\phi}} \cdot \frac{1}{[1-3\vartheta(1-\vartheta)]^{\frac{1}{2}}}. \tag{C.15}$$

However, if the magnitude of autocorrelation $\phi$ in Eq. C.14 has to be estimated, it is error prone and thus also Eq. C.14 is actually an approximation $\sigma_N^2 \approx \sigma_\epsilon^2/(1-\hat{\phi}^2)$.

More details on the estimation of the trend uncertainty can be found in Tiao et al. (1990) and Weatherhead et al. (1998).

# D Bayes' theorem

From the basic algebra of probability theory, the sum rule

$$P(Y|I) + P(\overline{Y}|I) = 1 \tag{D.1}$$

and product rule

$$P(Y,X|I) = P(Y|X,I) \cdot P(X|I), \tag{D.2}$$

Bayes' theorem can be derived as:

$$P(Y|X,I) = \frac{P(X|Y,I) \cdot P(Y,I)}{P(X|I)}, \tag{D.3}$$

where $X$ and $Y$ are propositions such as "it is cloudy" or "it is raining". $P(Y|X,I)$ is the probability of $Y$ conditional on $X$ and $I$, where $I$ denotes relevant background information. For example, we can ask for the conditional probability that it is raining given a cloudy sky: $P(raining|cloudy,I)$. $I$ could in this case be e.g. that we saw the weather forecast in the morning.

$P(Y|X,I)$ is called the posterior probability, $P(X|Y,I)$ is the likelihood, $P(Y,I)$ is the prior probability and $P(X|I)$ has former been called the marginalization likelihood, but nowadays Sivia and Skilling (2006) introduced the term 'evidence' for the denominator.

An important aspect from probability theory is the marginalization rule

$$P(X|I) = \int_{-\infty}^{+\infty} P(X,Y|I)\,dY = \int_{-\infty}^{+\infty} P(X|Y,I) \cdot P(Y|I)\,dY, \tag{D.4}$$

which can be well elucidated by an example taken from Sivia and Skilling (2006): Suppose there are $M$ candidates in a presidential election; then $Y_1$ could be the proposition that the first candidate will win, $Y_2$ that the second will win, and so on. Let $X$ be the proposition e.g. that unemployment will be lower the next year, irrespective of whoever becomes president, then:

$$P(X|I) = \sum_{k=1}^{M} P(X,Y_k|I) = \sum_{k=1}^{M} P(X|Y_k,I) \cdot P(Y_k|I). \tag{D.5}$$

The term $P(X|Y_k, I)$ in Eq. D.5 denotes the probability that unemployment will be reduced, **if** the candidate $Y_k$ has been chosen as president. This is multiplied by the probability that the candidate $Y_k$ will win the election, which is $P(Y_k|I)$. This product constitutes the joint probability that the unemployment will be reduced **and** the candidate $Y_k$ will win the election. Summing over all possibilities makes the result independent of whoever becomes president.

Equation D.4 represents the *continuum limit* of Eq. D.5 where we consider an arbitrarily large number of propositions.

Descriptive, Bayes' theorem (Eq. D.3) can be understood as:

$$P(hypothesis|data, I) \propto P(data|hypothesis, I) \cdot P(hypothesis, I). \tag{D.6}$$

Often the probability of the hypothesis given the data is hard to infer, whereas a better chance is given to estimate the probability of the data, if the hypothesis was true. This relation is exactly the power of Bayes' theorem.

An instructive example is given in Hütt and Dehnert (2006): A betimes unfair casino which makes use of two kinds of dice. One fair dice $W_1$ is used in 99% of the cases:

$$W_1: \quad P(W_1|I) = 0.99 \quad \text{and} \quad P(i|W_1, I) = \frac{1}{6}, \quad i = 1, ..., 6 \tag{D.7}$$

where the relevant background information, e.g. that the dice is used in 99% of the cases and that it is fair is absorbed in $I$. With a frequency of 1% an unfair dice is used:

$$\begin{aligned} W_2: \quad P(W_2|I) &= 0.01 \quad \text{and} \\ P(6|W_2, I) &= \frac{1}{2} \quad \text{and} \\ P(i|W_2, I) &= \frac{1 - P(6|W_2, I)}{5} = \frac{1}{10}, \quad i = 1, ..., 5 \end{aligned} \tag{D.8}$$

Now we suppose that we have chosen one of the dice and have thrown 3 times a six. Asking for the posterior probability that these three sixes have been thrown with the unfair dice $P(W_2|D, I)$ with the data $D = 6, 6, 6$ we can set up Bayes' theorem:

$$P(W_2|D, I) = \frac{P(D|W_2, I) \cdot P(W_2|I)}{P(D|I)}. \tag{D.9}$$

We can split up Eq. D.9 and calculate the terms separately: The prior probability is given above:

$$P(W_2|I) = 0.01. \tag{D.10}$$

The likelihood constitutes:

$$P(D|W_2, I) = \prod_{i=1}^{3} P(6|W_2, I) = \left(\frac{1}{2}\right)^3 = \frac{1}{8}. \tag{D.11}$$

The evidence for the data $D$, which is in fact a normalisation constant, is observed as:

$$\begin{align}
P(D|I) &= P(D, W_1|I) + P(D, W_2|I) \tag{D.12}\\
&= P(D|W_1, I) \cdot P(W_1|I) + P(D|W_2, I) \cdot P(W_2|I) \tag{D.13}\\
&= \left(\frac{1}{2}\right)^3 \cdot 0.01 + \left(\frac{1}{6}\right)^3 \cdot 0.99 \tag{D.14}\\
&= 0.005833. \tag{D.15}
\end{align}$$

Thus we can calculate the posterior probability:

$$\begin{align}
P(W_2|D, I) &= \frac{P(D|W_2, I) \cdot P(W_2|I)}{P(D|I)} \tag{D.16}\\
&= \frac{0.125 \cdot 0.01}{0.005833} \tag{D.17}\\
&= 0.21 \tag{D.18}
\end{align}$$

This result shows, that even if three sixes have been thrown in a row it is more likely that the fair dice has produced this series. This also demonstrates the importance of the prior probability.

# E Bayesian model selection

Bayesian model selection presents in principle the counterpart to the hypothesis testing in standard statistics. Here, the mathematical formalism of the model selection is shown, which can be applied to arbitrary models/hypotheses. We follow the very comprehensive description in Sivia and Skilling (2006), where we also show the derivation of the method for two hypotheses or models $A$ and $B$, which can easily be extended to more than two models.

The posterior probability of the hypothesis $A$, given the respective data $D$ using Bayes theorem is given by:

$$P(A|D,I) = \frac{P(D|A,I) \cdot P(A|I)}{P(D|I)}. \tag{E.1}$$

where $I$ describes certain relevant background information.

Usually the models are functions of certain parameters $a = a_1, a_2, ..., a_n$ and $b = b_1, b_2, ..., b_m$, which are often fitted to the data. Regarding the problem of model selection, the absolute magnitudes of the parameters $a$ and $b$ are mostly irrelevant. For instance, if the model selection is between a Gaussian distribution or a Lorentzian distribution, the position of the mean value does not influence the choice of the model.

Thus the marginalization rule Eq. D.4 can be used to eliminate the irrelevant parameters by integration:

$$P(A|D,I) = \frac{\int da\, P(D|A,a,I) \cdot P(A,a|I)}{P(D|I)} \tag{E.2}$$

Assuming logical independence of the prior probabilities of the hypothesis $A$ and the parameters $a$ we can simplify Eq. E.3, which yields:

$$P(A|D,I) = \frac{\int da\, P(D|A,a,I) \cdot P(a|I) \cdot P(A|I)}{P(D|I)} \tag{E.3}$$

In the same way the posterior for hypothesis B can be derived:

$$P(B|D,I) = \frac{\int db\, P(D|B,b,I) \cdot P(b|I) \cdot P(B|I)}{P(D|I)} \tag{E.4}$$

The denominator $P(D|I)$ is the evidence (cf. Eqs. D.4 and D.5), which is a normalisation constant and given by:

$$P(D|I) = P(D|A,I) \cdot P(A|I) + P(D|B,I) \cdot P(B|I). \tag{E.5}$$

Often the prior probabilities of the hypotheses $P(A|I)$ and $P(B|I)$ are chosen as equal, because none of them is preferred in the first place. In this case the priors of the hypotheses cancel out.

The parameter priors $P(a|I)$ and $P(b|I)$ are chosen, in many times, as bounded priors in the form of fully normalised uniform distribution:

$$P(p|I) = \begin{cases} \frac{1}{p_{max}-p_{min}} & \text{If } p_{min} < p < p_{max} \\ 0 & \text{otherwise} \end{cases} \quad p = a \text{ or } b. \tag{E.6}$$

The incorporation of prior probabilities is one major advantage of the Bayesian approach over the standard statistics. However if only very rare information about the parameters is known the choice of the priors is a hard task. The main requirement to the priors is, that they do not cut off any relevant probability space. Additionally, the priors should not be too large, because then they are quite uninformative and no longer advantageous.

The likelihoods $P(D|A,I)$ and $P(D|B,I)$ present the main part of the respective hypothesis or model. For instance, if we would hypothesise, that our data $D = D_1, ..., D_n$ follow a Gaussian normal distribution with parameters $a = (\mu, \sigma^2)$ we would use

$$P(D|A, \mu, \sigma, I) = \left(\sigma\sqrt{2\pi}\right)^{-N} \exp\left[-\frac{1}{2\sigma^2} \sum_{t=1}^{N} (D_t - \mu)^2\right], \tag{E.7}$$

as the likelihood for $A$. To calculate the posterior we have to integrate over $\mu$ and $\sigma$ in Eq. E.7.

In Bayesian model selection often the *Bayes factor* ($BF$) is used to decide between the models. The Bayes factor is the ratio:

$$BF = \frac{P(D|B,I) \cdot P(B|I)}{P(D|A,I) \cdot P(A|I)}. \tag{E.8}$$

As mentioned above mostly $P(A|I) = P(B|I)$, thus:

$$BF = \frac{P(D|B,I)}{P(D|A,I)} = \frac{\int db\, P(D|B,b,I)}{\int da\, P(D|A,a,I)}. \tag{E.9}$$

The Bayes factor gives the evidence against model $A$. Jeffreys (1939) has proposed a scale for this evidence, which is still used today as a guideline (cf. Tab.5.1).

Bayesian model selection is often used to decide between a simple and a more complex model. This is the case for instance, if certain data are observed and one wants to decide between fitting a polynomial of degree one or two to the data. In the sense of minimising the residuals the more complex model is mostly superior to the simple model, because it contains more parameters, which can be adjusted to fit the data. But regarding the problem of over-fitting, the common sense would mostly favour a polynomial of order one against e.g. a polynomial of order 10, although it has the smaller residuals. This circumstance is known as *Ockhams Razor*, a principle which recommends to choose the theory or model with the fewest assumptions and postulates when multiple competing theories are equal in describing respective phenomena. *Ockhams Razor* is naturally implemented in the Bayesian concept, in such a way, that a theory is penalised for every additional parameter automatically.

We can qualitatively derive the *Ockham factor* also shown in Sivia and Skilling (2006) and Dose and Menzel (2004). If model $B$ is e.g. the more complex polynomial of degree two and model $A$ is the simple polynomial of order one, there is one more dimension to integrate for $B$ denoted as $b_m$. This contribution to the integral is proportional to the width of the probability density function (pdf) $P(B|D,I)$ in this direction denoting as $\delta b_m$. With $P(b_m|I) = 1/(b_{m_{max}} - b_{m_{min}}) = 1/\triangle b_m$ we see, that the *Ockham factor* is $\approx \delta b_m/\triangle b_m$. This ratio is typically smaller than unity, thus penalises model $B$ for its additional parameter.

# F Analytical approximation – the matrices

Regarding Sect. 5.6, the quadratic Taylor series expansion of the logarithm of $A$'s likelihood function Eq. 5.20 yields:

$$L_A = L_A(\widehat{p}_1) - \frac{1}{2} K_A' H_A K_A + \cdots, \tag{F.1}$$

with

$$K_A' = p_1 - \widehat{p}_1 = \tag{F.2}$$
$$\begin{bmatrix} \mu_{p1} - \widehat{\mu}_{p1} & \mu_{p2} - \widehat{\mu}_{p2} & \omega_p - \widehat{\omega}_p & \delta_p - \widehat{\delta}_p \\ \sigma_{p1} - \widehat{\sigma}_{p1} & \sigma_{p2} - \widehat{\sigma}_{p2} \end{bmatrix}$$

and

$$H_A = \begin{pmatrix} \frac{\partial^2 L_A}{\partial \mu_{p1}^2} & \frac{\partial^2 L_A}{\partial \mu_{p1} \partial \mu_{p2}} & \frac{\partial^2 L_A}{\partial \mu_{p1} \partial \omega_p} & \frac{\partial^2 L_A}{\partial \mu_{p1} \partial \delta_p} & \frac{\partial^2 L_A}{\partial \mu_{p1} \partial \sigma_{p1}} & \frac{\partial^2 L_A}{\partial \mu_{p1} \partial \sigma_{p2}} \\ \frac{\partial^2 L_A}{\partial \mu_{p2} \partial \mu_{p1}} & \frac{\partial^2 L_A}{\partial \mu_{p2}^2} & \frac{\partial^2 L_A}{\partial \mu_{p2} \partial \omega_p} & \frac{\partial^2 L_A}{\partial \mu_{p2} \partial \delta_p} & \frac{\partial^2 L_A}{\partial \mu_{p2} \partial \sigma_{p1}} & \frac{\partial^2 L_A}{\partial \mu_{p2} \partial \sigma_{p2}} \\ \frac{\partial^2 L_A}{\partial \omega_p \partial \mu_{p1}} & \frac{\partial^2 L_A}{\partial \omega_p \partial \mu_{p2}} & \frac{\partial^2 L_A}{\partial \omega_p^2} & \frac{\partial^2 L_A}{\partial \omega_p \partial \delta_p} & \frac{\partial^2 L_A}{\partial \omega_p \partial \sigma_{p1}} & \frac{\partial^2 L_A}{\partial \omega_p \partial \sigma_{p2}} \\ \frac{\partial^2 L_A}{\partial \delta_p \partial \mu_{p1}} & \frac{\partial^2 L_A}{\partial \delta_p \partial \mu_{p2}} & \frac{\partial^2 L_A}{\partial \delta_p \partial \omega_p} & \frac{\partial^2 L_A}{\partial \delta_p^2} & \frac{\partial^2 L_A}{\partial \delta_p \partial \sigma_{p1}} & \frac{\partial^2 L_A}{\partial \delta_p \partial \sigma_{p2}} \\ \frac{\partial^2 L_A}{\partial \sigma_{p1} \partial \mu_{p1}} & \frac{\partial^2 L_A}{\partial \sigma_{p1} \partial \mu_{p2}} & \frac{\partial^2 L_A}{\partial \sigma_{p1} \partial \omega_p} & \frac{\partial^2 L_A}{\partial \sigma_{p1} \partial \delta_p} & \frac{\partial^2 L_A}{\partial \sigma_{p1}^2} & \frac{\partial^2 L_A}{\partial \sigma_{p1} \partial \sigma_{p2}} \\ \frac{\partial^2 L_A}{\partial \sigma_{p2} \partial \mu_{p1}} & \frac{\partial^2 L_A}{\partial \sigma_{p2} \partial \mu_{p2}} & \frac{\partial^2 L_A}{\partial \sigma_{p2} \partial \omega_p} & \frac{\partial^2 L_A}{\partial \sigma_{p2} \partial \delta_p} & \frac{\partial^2 L_A}{\partial \sigma_{p2} \partial \sigma_{p1}} & \frac{\partial^2 L_A}{\partial \sigma_{p2}^2} \end{pmatrix}, \tag{F.3}$$

which is

$$H_A = \begin{pmatrix} \frac{\ell_1}{\widehat{\sigma}_{p1}^2} & 0 & \frac{\sum X_{1t}}{\widehat{\sigma}_{p1}^2} & \frac{\sum U_t}{\widehat{\sigma}_{p1}^2} & \frac{2}{\widehat{\sigma}_{p1}} & 0 \\ 0 & \frac{\ell_2}{\widehat{\sigma}_{p2}^2} & \frac{\sum X_{2t}}{\widehat{\sigma}_{p2}^2} & 0 & 0 & \frac{2}{\widehat{\sigma}_{p2}} \\ \frac{\sum X_{1t}}{\widehat{\sigma}_{p1}^2} & \frac{\sum X_{2t}}{\widehat{\sigma}_{p2}^2} & \frac{\sum X_{1t}^2}{\widehat{\sigma}_{p1}^2} + \frac{\sum X_{2t}^2}{\widehat{\sigma}_{p2}^2} & \frac{\sum U_t X_{1t}}{\widehat{\sigma}_{p1}^2} & \frac{2 \sum \epsilon_{p1t} X_{1t}}{\widehat{\sigma}_{p1}^3} & \frac{2 \sum \epsilon_{p2t} X_{2t}}{\widehat{\sigma}_{p2}^3} \\ \frac{\sum U_t}{\widehat{\sigma}_{p1}^2} & 0 & \frac{\sum X_{1t} U_t}{\widehat{\sigma}_{p1}^2} & \frac{\sum U_t^2}{\widehat{\sigma}_{p1}^2} & \frac{2 \sum \epsilon_{p1t} U_t}{\widehat{\sigma}_{p1}^3} & 0 \\ \frac{2}{\widehat{\sigma}_{p1}} & 0 & \frac{2 \sum \epsilon_{p1t} X_{1t}}{\widehat{\sigma}_{p1}^3} & \frac{2 \sum \epsilon_{p1t} U_t}{\widehat{\sigma}_{p1}^3} & \frac{2\ell_1}{\widehat{\sigma}_{p1}^2} & 0 \\ 0 & \frac{2}{\widehat{\sigma}_{p2}^2} & \frac{2 \sum \epsilon_{p2t} X_{2t}}{\widehat{\sigma}_{p2}^3} & 0 & 0 & \frac{2\ell_2}{\widehat{\sigma}_{p2}^2} \end{pmatrix} \tag{F.4}$$

Analogues the quadratic Taylor series expansion of the logarithm of $B$'s likelihood function Eq. 5.21 yields:

$$L_B = L_B(\widehat{p}_2) - \frac{1}{2} K_B' \mathbf{H}_B K_B + \cdots, \tag{F.5}$$

with

$$K_B' = p_2 - \widehat{p}_2 = \tag{F.6}$$
$$\begin{bmatrix} \mu_1 - \widehat{\mu}_1 & \mu_2 - \widehat{\mu}_2 & \omega_1 - \widehat{\omega}_1 & \omega_2 - \widehat{\omega}_2 & \delta - \widehat{\delta} \\ \sigma_1 - \widehat{\sigma}_1 & \sigma_2 - \widehat{\sigma}_2 \end{bmatrix}$$

and

$$\mathbf{H}_B = \begin{pmatrix} \frac{\partial^2 L_B}{\partial \mu_1^2} & \frac{\partial^2 L_B}{\partial \mu_1 \partial \mu_2} & \frac{\partial^2 L_B}{\partial \mu_1 \partial \omega_1} & \frac{\partial^2 L_B}{\partial \mu_1 \partial \omega_2} & \frac{\partial^2 L_B}{\partial \mu_1 \partial \delta} & \frac{\partial^2 L_B}{\partial \mu_1 \partial \sigma_1} & \frac{\partial^2 L_B}{\partial \mu_1 \partial \sigma_2} \\ \frac{\partial^2 L_B}{\partial \mu_2 \partial \mu_1} & \frac{\partial^2 L_B}{\partial \mu_2^2} & \frac{\partial^2 L_B}{\partial \mu_2 \partial \omega_1} & \frac{\partial^2 L_B}{\partial \mu_2 \partial \omega_2} & \frac{\partial^2 L_B}{\partial \mu_2 \partial \delta} & \frac{\partial^2 L_B}{\partial \mu_2 \partial \sigma_1} & \frac{\partial^2 L_B}{\partial \mu_2 \partial \sigma_2} \\ \frac{\partial^2 L_B}{\partial \omega_1 \partial \mu_1} & \frac{\partial^2 L_B}{\partial \omega_1 \partial \mu_2} & \frac{\partial^2 L_B}{\partial \omega_1^2} & \frac{\partial^2 L_B}{\partial \omega_1 \partial \omega_2} & \frac{\partial^2 L_B}{\partial \omega_1 \partial \delta} & \frac{\partial^2 L_B}{\partial \omega_1 \partial \sigma_1} & \frac{\partial^2 L_B}{\partial \omega_1 \partial \sigma_2} \\ \frac{\partial^2 L_B}{\partial \omega_2 \partial \mu_1} & \frac{\partial^2 L_B}{\partial \omega_2 \partial \mu_2} & \frac{\partial^2 L_B}{\partial \omega_2 \partial \omega_1} & \frac{\partial^2 L_B}{\partial \omega_2^2} & \frac{\partial^2 L_B}{\partial \omega_2 \partial \delta} & \frac{\partial^2 L_B}{\partial \omega_2 \partial \sigma_1} & \frac{\partial^2 L_B}{\partial \omega_2 \partial \sigma_2} \\ \frac{\partial^2 L_B}{\partial \delta \partial \mu_1} & \frac{\partial^2 L_B}{\partial \delta \partial \mu_2} & \frac{\partial^2 L_B}{\partial \delta \partial \omega_1} & \frac{\partial^2 L_B}{\partial \delta \partial \omega_2} & \frac{\partial^2 L_B}{\partial \delta^2} & \frac{\partial^2 L_B}{\partial \delta \partial \sigma_1} & \frac{\partial^2 L_B}{\partial \delta \partial \sigma_2} \\ \frac{\partial^2 L_B}{\partial \sigma_1 \partial \mu_1} & \frac{\partial^2 L_B}{\partial \sigma_1 \partial \mu_2} & \frac{\partial^2 L_B}{\partial \sigma_1 \partial \omega_1} & \frac{\partial^2 L_B}{\partial \sigma_1 \partial \omega_2} & \frac{\partial^2 L_B}{\partial \sigma_1 \partial \delta} & \frac{\partial^2 L_B}{\partial \sigma_1^2} & \frac{\partial^2 L_B}{\partial \sigma_1 \partial \sigma_2} \\ \frac{\partial^2 L_B}{\partial \sigma_2 \partial \mu_1} & \frac{\partial^2 L_B}{\partial \sigma_2 \partial \mu_2} & \frac{\partial^2 L_B}{\partial \sigma_2 \partial \omega_1} & \frac{\partial^2 L_B}{\partial \sigma_2 \partial \omega_2} & \frac{\partial^2 L_B}{\partial \sigma_2 \partial \delta} & \frac{\partial^2 L_B}{\partial \sigma_2 \partial \sigma_1} & \frac{\partial^2 L_B}{\partial \sigma_2^2} \end{pmatrix}, \tag{F.7}$$

which is

$$\mathbf{H}_B = \begin{pmatrix} \frac{\ell_1}{\widehat{\sigma}_1^2} & 0 & \frac{\sum X_{1t}}{\widehat{\sigma}_1^2} & 0 & \frac{\sum U_t}{\widehat{\sigma}_1^2} & \frac{2}{\widehat{\sigma}_1} & 0 \\ 0 & \frac{\ell_2}{\widehat{\sigma}_2^2} & 0 & \frac{\sum X_{2t}}{\widehat{\sigma}_2^2} & 0 & 0 & \frac{2}{\widehat{\sigma}_2} \\ \frac{\sum X_{1t}}{\widehat{\sigma}_1^2} & 0 & \frac{\sum X_{1t}^2}{\widehat{\sigma}_1^2} & 0 & \frac{\sum U_t X_{1t}}{\widehat{\sigma}_1^2} & \frac{2\sum \epsilon_{1t} X_{1t}}{\widehat{\sigma}_1^3} & 0 \\ 0 & \frac{\sum X_{2t}}{\widehat{\sigma}_2^2} & 0 & \frac{\sum X_{2t}^2}{\widehat{\sigma}_2^2} & 0 & 0 & \frac{2\sum \epsilon_{2t} X_{2t}}{\widehat{\sigma}_2^3} \\ \frac{\sum U_t}{\widehat{\sigma}_1^2} & 0 & \frac{\sum X_{1t} U_t}{\widehat{\sigma}_1^2} & 0 & \frac{\sum U_t^2}{\widehat{\sigma}_1^2} & \frac{2\sum \epsilon_{1t} U_t}{\widehat{\sigma}_1^3} & 0 \\ \frac{2}{\widehat{\sigma}_1} & 0 & \frac{2\sum \epsilon_{1t} X_{1t}}{\widehat{\sigma}_1^3} & 0 & \frac{2\sum \epsilon_{1t} U_t}{\widehat{\sigma}_1^3} & \frac{2\ell_1}{\widehat{\sigma}_1^2} & 0 \\ 0 & \frac{2}{\widehat{\sigma}_2} & 0 & \frac{2\sum \epsilon_{2t} X_{2t}}{\widehat{\sigma}_2^3} & 0 & 0 & \frac{2\ell_2}{\widehat{\sigma}_2^2} \end{pmatrix} \tag{F.8}$$

# Bibliography

Andersson, A., Bakan, S., Fennig, K., H., Klepp, C. P., and Schulz, J.: Atmosphere Parameters and Fluxes from Satellite Data - HOAPS - 3 monthly mean, doi:10.1594/WDCC/HOAPS3_MONTHLY],, 2007.

Bayes, T.: An essay towards solving a problem in the doctrine of chances, Phil.Trans.Roy.Soc., 53, 330–418, 1763.

Berger, J. O.: Statistical Decision Theory and Bayesian Analysis, Springer, 1993.

Boucher, O., Myhre, G., and Myhre, A.: Direct human influence of irrigation on atmospheric water vapour and climate, Climate Dynamics, 22, 597–603, 2004.

Bovensmann, H., Burrows, J. P., Buchwitz, M., Frerick, J., Noël, S., Rozanov, V. V., Chance, K. V., and Goede, A. H. P.: SCIAMACHY–Mission objectives and measurement modes, J. Atmos. Sci., 56, 127–150, 1999.

Braak, C. T.: A Markov Chain Monte Carlo version of the genetic algorithm Differential Evolution: easy Bayesian computing for real parameter spaces, Stat. Comput., 16, 239–249, 2006.

Bretthorst, G. L.: On The Difference In Means, In Physics and Probability. Cambridge University Press, pp. 177–194, 1993.

Brocard, E.: Overview on satellite experiments which measure atmospheric water vapor, IAP Research Report, Institut für angewandte Physik, Universität Bern, Bern, Switzerland, 2006.

Buchwitz, M., de Beek, R., Noël, S., Burrows, J. P., Bovensmann, H., Schneising, O., Khlystova, I., Bruns, M., Bremer, H., Bergamaschi, P., Körner, S., and Heimann, M.: Atmospheric carbon gases retrieved from SCIAMACHY by WFM-DOAS: version 0.5 CO and CH4 and impact of calibration improvements on CO2 retrieval, Atmos. Chem. Phys., 6, 2727–2751, 2006.

Burrows, J. P., Schneider, W., Geary, J. C., Chance, K. V., Goede, A. P. H., Aarts, H. J. M., de Vries, J., Smorenburg, C., and Visser, H.: Atmospheric remote sensing with SCIAMACHY, Digest of Topical Meeting on Optical Remote Sensing of the Atmosphere, Optical Society of America, Washington, 4, 71–74, 1990.

Burrows, J. P., Hölzle, E., Goede, A. P. H., Visser, H., and Fricke, W.: SCIAMACHY - Scanning imaging absorption spectrometer for atmospheric chartography, Acta Astronautica, 35, 445–451, 1995.

Burrows, J. P., Weber, M., Buchwitz, M., Rozanov, V., Ladstätter-Weißenmayer, A., Richter, A., de Beek, R., Hoogen, R., Bramstedt, K., Eichmann, K.-U., Eisinger, M., and Perner, D.: The Global Ozone Monitoring Experiment (GOME): Mission Concept and First Scientific Results, J. Atmos. Sci., 56, 151–175, 1999.

Dai, A., Meehl, G. A., Washington, W. M., Wigley, T. M. L., and Arblaster, J. A.: Ensemble simulation of twenty-first century climate changes: Business-as-usual versus CO2 stabilization, Bull. Amer. Meteor. Soc., 82, 2377–2388, 2001.

de Laplace, P. S.: Théorie analytique des probalités, Courcier Imprimeur, Paris, 1812.

Diamond, J.: Collaps. How Societies Choose to Fail or Succeed, Viking, Penguin Group, New York, 2005.

Dose, V. and Menzel, A.: Bayesian analysis of climate change impacts in phenology, Global Change Biology, 10, 259–272, 2004.

Dose, V. and Menzel, A.: Bayesian correlation between temperature and blossom onset data, Global Change Biology, 12, 1451–1459, 2006.

Ebeling, W., Freund, J., and Schweitzer, F.: Komplexe Strukturen: Entropie und Information, B. G. Teubner, 1998.

Edelson, R. A. and Krolik, J. H.: The discrete correlation function: A new method for analyzing unevenly sampled variability data, The Astrophysical Journal, 333, 646–659, 1988.

EEA/JRC/WHO: Impacts of Europe's changing climate – 2008 indicator-based assessment, EEA Report, 4, http://reports.eea.europa.eu/eea_report_2008_4/en, 2008.

Fahrmeir, L., Pigeot, I., Künstler, R., and Tutz, G.: Statistik, Springer, Berlin, 2004.

Freund, J. A.: Dynamische Entropien und nichtlineare Prozesse mit langreichweitigen Korrelationen, Dissertation, Humboldt Universität Berlin, 1996.

Freund, J. A., Pöschel, T., and Wiltshire, K. H.: Markovsche Analyse nasser Gemeinschaften, pp. 99–110, Logos, Berlin, 2006.

Gilks, W. R., Richardson, S., and Spiegelhalter, D.: Markov Chain Monte Carlo in Practice, Chapman & Hall/CRC, 1995.

Good, S. A., Corlett, G. K., Remedios, J. J., Noyes, E. J., and Llewellyn-Jones, D. T.: The Global Trend in Sea Surface Temperature from 20 Years of Advanced Very High Resolution Radiometer Data, Journal of Climate, 20, 1255–1264, 2007.

Gordon, L. J., Steffen, W., Jörnsen, B. F., Folke, C., Falkenmark, M., and Johannessen, Å.: Human modification of global water vapor flows from the land surfaces, PNAS, 102, 7612–7617, 2005.

Gottwald, M., Bovensmann, H., Lichtenberg, G., Noël, S., von Bargen, A., Slijkhuis, S., Piters, A., Hoogeveen, R., von Savigny, C., Buchwitz, M., Kokhanovsky, A., Richter, A., Rozanov, A., Holzer-Popp, T., Bramstedt, K., Lambert, J.-C., Skupin, J., Wittrock, F., Schrijver, H., and Burrows, J.: SCIAMACHY, Monitoring the Changing Earth's Atmosphere, DLR, 2006.

Granger, C. W. J.: Investigating causal relations by econometric models and cross-spectral methods, Econometrica, 37, 424–438, 1969.

Granger, C. W. J.: Essays in econometrics: The collected papers of clive w.j. granger, Cambridge University Press, p. 310ff, 2001.

Häckel, H.: Meteorologie, Ulmer, Stuttgart, Germany, 1999.

Hansen, J. E. and Lebedeff, S.: Global trends of measured surface air temperature, J. Geophys. Res., 92, 1992.

Held, I. M. and Soden, B. J.: Water Vapor Feedback And Global Warming, Annu. Rev. Energy Environ., 25, 441–75, 2000.

Hill, M. F., Witman, J. D., and Caswell, H.: Markov Chain Analysis of Succession in a Rocky Subtidal Community, The American Naturalist, 164, E46–E61, 2004.

Horn, R. A. and Johnson, C. R.: Matrix Analysis, Cambridge University Press, 1990.

Hütt, M.-T. and Dehnert, M.: Methoden der Bioinformatik, Springer, 2006.

IPCC: Climate Change 2007: The Physical Science Basis. Contribution of Working Group I to the Fourth Assessment Report of the Intergovernmental Panel on Climate Change, S. Solomon, D. Qin, M. Manning, Z. Chen, M. Marquis, K.B. Averyt, M. Tignor and H.L. Miller, eds., Cambridge University Press, Cambridge, United Kingdom and New York, NY, USA, 996 pp., 2007.

Isagi, Y. and Nakagoshi, N.: A Markov approach for describing post-fire succession of vegetation, Ecological Research, 5, 163–171, 1990.

Jaynes, E. T. and Bretthorst, L. G.: Probability Theory: The Logic of Science: Principles and Elementary Applications Vol 1, Cambridge University Press, Cambridge, 2003.

Jeffreys, H.: Theory of Probability, Oxford University Press, USA; 3 edition (November 12, 1998), 1939.

Kac, M.: On the notion of recurrence in discrete stochastic processes, Bull. Amer. Math. Soc, 53, 1002–1010, 1947.

Kaufmann, R. K. and Stern, D. I.: Evidence for human influence on climate from hemispheric temperature relations, Nature, 388, 39–44, 1997.

Lang, R., Williams, J. E., van der Zande, W. J., and Maurellis, A. N.: Application of the Spectral Structure Parameterization technique: retrieval of total water vapor columns from GOME, Atmos. Chem. Phys., 3, 145–160, 2003.

Lanzante, J. R.: A Cautionary Note on the Use of Error Bars, Journal of Climate, 18, 3699–3703, 2005.

Lenderink, G. and Meijgaard, E. V.: Increase in hourly precipitation extremes beyond expectations from temperature changes, Nature Geoscience, 1, 511–514, 2008.

Liu, J. S.: Monte Carlo Strategies in Scientific Computing, Springer, New York, 2003.

Lorenz, E. N.: Deterministic Nonperiodic Flow, J. Atmos. Sci., 20, 130–141, 1963.

Markov, A. A.: Wahrscheinlichkeitsrechnung, Teubner, Berlin, 1912.

Maurellis, A. N., Lang, R., van der Zande, W. J., Aben, I., and Ubachs, W.: Precipitable water vapor column retrieval from GOME data, Geophys. Res. Lett., 27, 903–906, 2000.

Melillo, J. M.: Climate Change: Warm, Warm on the Range, Science, 8, 183–184, 1999.

Mieruch, S., Noël, S., Bovensmann, H., and Burrows, J. P.: Analysis of global water vapour trends from satellite measurements in the visible spectral range, Atmos. Chem. Phys., 8, 491–504, 2008.

Moon, S.-E., Ryoo, S.-B., and Kwon, J.-G.: A markov chain model for daily precipitation occurence in south korea, Int. J. Climatol., 14, 1009–1016, 1994.

Nicolis, C., Ebeling, W., and Baraldi, C.: Markov processes, dynamic entropies and the statistical prediction of mesoscale wheather regimes, Tellus A, 49, 108–118, 1997.

Noël, S., Buchwitz, M., Bovensmann, H., Hoogen, R., and Burrows, J. P.: Atmospheric Water Vapor Amounts Retrieved from GOME Satellite Data, Geophys. Res. Lett., 26(13), 1841–1844, 1999.

Noël, S., Buchwitz, M., and Burrows, J. P.: First retrieval of global water vapour column amounts from SCIAMACHY measurements, Atmos. Chem. Phys., 4, 111–125, 2004.

Noël, S., Buchwitz, M., Bovensmann, H., and Burrows, J. P.: Validation of SCIAMACHY AMC-DOAS water vapour columns, Atmos. Chem. Phys., 5, 1835–1841, 2005.

Noël, S., Mieruch, S., Buchwitz, M., Bovensmann, H., and Burrows, J. P.: GOME and SCIAMACHY global water vapour columns, in Proceedings of the First Atmospheric Science Conference, ESA Publications Devision, Noordwijk, The Netherlands, http://earth.esa.int/cgi-bin/confatmos06.pl?abstract=166, 2006.

Noël, S., Mieruch, S., Bovensmann, H., and Burrows, J. P.: A combined GOME and SCIAMACHY global water vapour data set (submitted), in ENVISAT Sumposium 2007, SP_636_ENVISAT, ESA Publications Devision, Noordwijk, The Netherlands, 2007.

Noël, S., Mieruch, S., Bovensmann, H., and Burrows, J. P.: Preliminary results of GOME-2 water vapour retrievals and first application in polar regions, Atmos. Chem. Phys., 8, 1519–1529, 2008.

Prölss, G. W.: Physik des erdnahen Weltraums, Springer, 2001.

Robert, C. P. and Casella, G.: Monte Carlo Statistical Methods, Springer, 2005.

Roedel, W.: Physik unserer Umwelt Die Atmosphäre, Springer-Verlag, 2000.

Scheffer, M. and Carpenter, S. R.: Catastrophic regime shifts in ecosystems: linking theory to observation, Trends Ecol. Evol., 18, 648–656, 2003.

Schlittgen, R. and Streitberg, B. H. J.: Zeitreihenanalyse, Oldenbourg, München, 1997.

Schneising, O., Buchwitz, M., Burrows, J. P., Bovensmann, H., Reuter, M., Notholt, J., Macatangay, R., and Warneke, T.: Three years of greenhouse gas column-averaged dry air mole fractions retrieved from satellite âÄŞ Part 1: Carbon dioxide, Atmos. Chem. Phys., 8, 3827–3853, 2008.

Schwarz, G.: Estimating the dimension of a model, Annals of Statistics, 6, 461–464, 1978.

Shannon, C. E.: A Mathematical Theory of Communication, Bell System Technical Journal, 27, 623–656, 1948.

Sivia, D. S. and Skilling, J.: Data Analysis A Bayesian Tutorial, Oxford University Press, 2006.

Stenke, A. and Grewe, V.: Simulation of stratospheric water vapor trends: impact on stratospheric ozone chemistry, Atmos. Chem. Phys., 5, 1257–1272, 2005.

Stoorvogel, J. J. and Fresco, L. O.: Quantification of land-use dynamics: An illustration from Costa Rica, Land Degradation and Development, 7, 121–131, 1996.

Storch, H. V. and Zwiers, F. W.: Statistical Analysis in Climate Research, Cambridge Universty Press, Cambridge, UK, 1999.

Tanner, J. E., Hughes, T. P., and Connell, J. H.: Species coexistence, keystone species, and succession: a sensitivity analysis, Ecology, 75, 2204–2219, 1994.

Thornton, D. L. and Batten, D. S.: Lag length selection and Granger causality, Working Paper 1984-001A, Federal Reserve Bank of St. Louis, http://research.stlouisfed.org/wp/1984/1984-001.pdf, 1984.

Tiao, G. C., Reinsel, G. C., Xu, D., Pedrick, J. H., Zhu, X., Miller, A. J., DeLuisi, J. J., Mateer, C. L., and Wuebbles, D. J.: Effects of autocorrelation and temporal sampling schemes on estimation of trend and spatial correlation, Journal of Geophysical Research, 95, 20,507–20,517, 1990.

Triacca, U.: Is Granger causality analysis appropriate to investigate the relationship between atmospheric concentration of carbon dioxide and global surface air temperature?, Theor. Appl. Climatol., 81, 133–135, 2005.

Turner, D. D., Lesht, B. M., Clough, S. A., Liljegren, J. C., Revercomb, H. E., and Tobin, D. C.: Dry bias and variability in Vaisala RS80-H radiosondes: The ARM Experiment, Journal of Atmospheric and Oceanic Technology, 20, 117–132, 2003.

Usher, M. B.: Markovian approaches to ecological succession, Journal of Animal Ecology, 48, 413–426, 1979.

Waggoner, P. E. and Stephens, G. R.: Transition probabilities for a forest, Nature, 255, 1160–1161, 1970.

Wagner, T., Heland, J., Zöger, M., and Platt, U.: A fast $H_2O$ total column density product from GOME – Validation with in-situ aircraft measurements, Atmos. Chem Phys, 3, 651–663, 2003.

Wagner, T., Beirle, S., Grzegorski, M., Sanghavi, S., and Platt, U.: El niño induced anomalies in global data sets of total column precipitable water and cloud cover derived from GOME on ERS-2, Journal of Geophysical Research, 110, D15 104, 2005.

Wagner, T., Beirle, S., Grzegorski, M., and Platt, U.: Global trends (1996-2003) of total column precipitable water observed by Global Ozone Monitoring Experiment (GOME on ERS-2) and their relation to near-surface temperature, Journal of Geophysical Research, 111, D12 102, 2006.

Waldmann, K.-H. and Stocker, U. M.: Stochastische Modelle, Springer, Berlin, 2003.

Weatherhead, E. C., Reinsel, G. C., Tiao, G. C., Meng, X.-L., Choi, D., Cheang, W.-K., Keller, T., DeLuisi, J., Wuebbles, D. J., Kerr, J. B., Miller, A. J., Oltmans, S. J., and Frederick, J. E.: Factors affecting the detection of trends: Statistical considerations and applications to environmental data, Journal of Geophysical Research, 103, 17,149–17,161, 1998.

Welch, B. L.: The generalization of "student's" problem when several different population variances are involved, Biometrika, 34, 28–35, 1947.

Wessel, N., Ziehmann, C., Kurths, J., Meyerfeldt, U., Schirdewan, A., and Voss, A.: Short-term forecasting of life-theatening cardiac arrhythmias based on symbolic dynamics and finite-time growth rates, Phys.Rev.E, 61, 733–739, 2000.

# Acknowledgements

I would like to thank the following people and institutions for their help and contribution throughout this project.

First of all I want to thank Prof. Dr. John P. Burrows for giving me the opportunity to work in his group at the IUP (Institute of Environmental Physics) on the fascinating subject of atmospheric science. I have appreciated his support and motivation during the last three years.

Special thanks go to my supervisor Stefan Noël who shared his expertise and research insight with me. Thanks for the fruitful discussions and valuable advice. Particularly, I am deeply thankful for his endeavours in struggling through my texts. Without his exceptional support, this work would not have been possible.

Further I would like to thank my group leader Heinrich Bovensmann. He always gave me a lot of support and motivation.

I would like to acknowledge the financial support from the European Space Agency (ESA) (GMES project PROMOTE), the German Aerospace Center (DLR) (project SADOS) and the University and the State of Bremen.

I thank the PIP (Postgraduate International Physics Programme) for financial benefit, allowing the participation at the AGU conference in Hawaii.

As a listing might go beyond the scope of these acknowledgements I say thanks to all my colleagues at the IUP. However, I especially thank Maximilian Reuter, Oliver Schneising and Ralf Bauer for fruitful discussions not only on Bayesian statistics.

Many thanks go to Jörg Schulz and Marc Schröder from the German Weather Service (DWD) for providing the radiosonde data, the Goddard Institute of Space Studies (GISS) for providing temperature data, Veronika Eyring from the DLR for fruitful discussions and Ronny Leinweber from the Freie Universität Berlin for information on radiosonde measurements.

I would also like to thank Jan Freund from the ICBM (Institute for Chemistry and Biology of the Marine Environment) in Oldenburg for valuable discussions from Bayesian concepts to Markov chains for more than four years now.

I am deeply grateful to my wife Kirsten for being with me for eight years. I thank her for her support, motivation and also for professional discussions with her as a physicist.

Zu guter Letzt möchte ich meinen Eltern danken, die mir das Studium ermöglicht haben und die immer für mich da waren und sind.

Die VDM Verlagsservicegesellschaft sucht für wissenschaftliche Verlage abgeschlossene und herausragende

## Dissertationen, Habilitationen, Diplomarbeiten, Master Theses, Magisterarbeiten usw.

für die kostenlose Publikation als Fachbuch.

Sie verfügen über eine Arbeit, die hohen inhaltlichen und formalen Ansprüchen genügt, und haben Interesse an einer honorarvergüteten Publikation?

Dann senden Sie bitte erste Informationen über sich und Ihre Arbeit per Email an *info@vdm-vsg.de*.

**Sie erhalten kurzfristig unser Feedback!**

VDM Verlagsservicegesellschaft mbH
Dudweiler Landstr. 99   Telefon +49 681 3720 174
D - 66123 Saarbrücken   Fax     +49 681 3720 1749
**www.vdm-vsg.de**

Die VDM Verlagsservicegesellschaft mbH vertritt

Printed by Books on Demand GmbH, Norderstedt / Germany